Mechanical Engineering Series

Frederick F. Ling
Editor-in-Chief

Mechanical Engineering Series

(continued after index)

Weili Cheng and Iain Finnie

Residual Stress Measurement and the Slitting Method

 Springer

Weili Cheng
Berkeley Engineering and Research, Inc.
Berkeley, CA
USA

Iain Finnie
Department of Mechanical Engineering
University of California, Berkeley
Berkeley, CA
USA

Residual Stress Measurement and the Slitting Method

Library of Congress Control Number: 2006931571

ISBN 0-387-37065-X   e-ISBN 0-387-39030-8
ISBN 978-0-387-37065-1   e-ISBN 978-0-387-39030-7

Printed on acid-free paper.

Printed in the United States of America.

9 8 7 6 5 4 3 2 1

springer.com

Mechanical Engineering Series

Frederick F. Ling
Editor-in-Chief

The Mechanical Engineering Series features graduate texts and research monographs to address the need for information in contemporary mechanical engineering, including areas of concentration of applied mechanics, biomechanics, computational mechanics, dynamical systems and control, energetics, mechanics of materials, processing, production systems, thermal science, and tribology.

Series Preface

Mechanical engineering, and engineering discipline born of the needs of the industrial revolution, is once again asked to do its substantial share in the call for industrial renewal. The general call is urgent as we face profound issues of productivity and competitiveness that require engineering solutions, among others. The Mechanical Engineering Series is a series featuring graduate texts and research monographs intended to address the need for information in contemporary areas of mechanical engineering.

The series is conceived as a comprehensive one that covers a broad range of concentrations important to mechanical engineering graduate education and research. We are fortunate to have a distinguished roster of consulting editors, each an expert in one of the areas of concentration. The names of the consulting editors are listed on page vi of this volume. The areas of concentration are applied mechanics, biomechanics, computational mechanics, dynamic systems and control, energetics, mechanics of materials, processing, thermal science, and tribology.

To
Weihsun and Joan

Preface

The early development of the slitting method is closely related to the work in Fracture Mechanics. G. Irwin's strain energy release rate is the foundation for computation of the crack compliance functions. H. F. Bueckner's principle for crack growth leads to the superposition principle for the release of the residual stresses. R. J. Hartrenft and G. G. Sih's, and G. Chell's expressions of K_I for shallow and deep cracks lead to an expression of K_I that works for both cases. The body force method introduced by H. Nisitani and his colleagues is a very useful tool for computing the compliance functions for a cut of finite width for near-surface measurements. The inherent-strain method developed by Y. Ueda and his colleagues has inspired the use of initial strains to approximate the residual stresses in the slitting method and the single-slice method.

The method of measuring residual stresses by a cut of progressively increasing depth was first tried in 1971 by S. Vaidyanathan and I. Finnie, who estimated the residual stress from a variation of K_I obtained by a photoelastic technique. It was not until fourteen years later that the method was extended by W. Cheng and I. Finnie in 1985 to measure residual stresses from a strain variation. In the years that followed a number of researchers around the world carried out similar measurements: D. Ritchie and R. H. Leggatt in 1987, T. Frett in 1987, C. N. Reid in 1988, and K. J. Kang, J. H. Song and Y. Y. Earmme in 1989.

Part of our early work was supported by Joe Gilman and Raj Pathania of EPRI and Wayne Kroenke of Bettis Laboratory. We appreciate the significant contribution of Mike Prime, who chairs the ASTM E.28.13.02 Task Group, and all Task Group members, Mike Hill, Gary Schajer, Yung Fan, Hans Schindler, and Can Aydiner, who have devoted their time to working towards a standard for the slitting method.

Our thanks also go to Öktam Vardar, Marco Gremaud, Glen Stevick, Ron Streit and Robert Ritchie for their contributions, friendship, and help over the years.

Fremont, California
Berkeley, California

Weili Cheng
Iain Finnie
June 2006

Contents

1

Introduction to Residual Stresses

1.1 What are residual stresses?

Residual stresses have been associated with humans ever since civilization began. The making of intricate clay components using fire in early days was actually an art that maintained the balance between reducing the residual stress gradient and achieving the desired shape of products. A stronger sword was often the result of a thin layer of compressive residual stress induced by repeated hammering at a controlled elevated temperature. Even today, the presence of residual stresses still dictates the design of many components, whether in a spacecraft or a tiny integrated circuit.

So what are residual stresses? In short, residual stresses are stress fields that exist in the absence of any external loads. All mechanical processes can cause deformation that may lead to residual stresses. For example, nonuniform heating or cooling causes thermal strains, plastic deformation induces incompatible deformation, and mismatched thermal expansion coefficients produce discontinuity in deformation under temperature change. Thus, the state of a residual stress depends on both the prior processes it has undergone, and the material properties that relate the current mechanical process/environment to deformation.

Figure 1.1 illustrates a thermal switch that makes use of residual stresses to produce desired movements. The switch arm is made of two layers with different thermal expansion coefficients, α_1 and α_2. At room temperature the length of the layers is the same and the arm is straight. Assuming $\alpha_1 > \alpha_2$, a temperature increase ΔT will make layer 1 expand more than layer 2. However, the bonding between the layers restricts layer 1 from expanding freely. The restriction can be visualized as a tension on layer 2 and a compression on layer 1, which are always in perfect balance. The resulting deformation controlled by the residual stress is an arm curved towards terminal 2. Similarly, a temperature drop $-\Delta T$ leads to an arm curved towards terminal 1. This simple example shows two fundamental features of residual stress: any

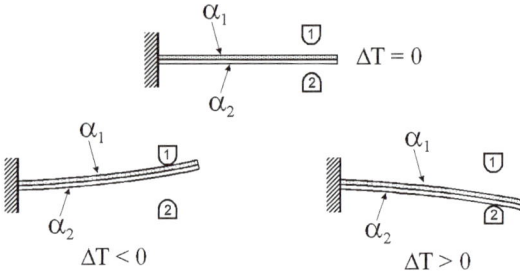

Fig. 1.1. A thermal switch that turns on/off depending on the temperature change.

tension/compression is always balanced by compression/tension, and the stress in the region restricted from expansion is compressive, and *vice versa*.

The study of residual stresses ranges from such common applications as the stresses existing in a bolted assembly to the special surface treatment by laser beams. Prediction and measurement of residual stresses in engineering components have been a constant pursuit of many researchers throughout the improvement of old or development of new components.

1.2 Influence of Residual Stresses on the Integrity of Mechanical Components

The mysterious cracking of a standing clay vase and a sudden tremor of the earth we live on have one thing in common: the releases of residual stresses cause deformation, whether it is an almost unnoticeable increment of cracking or a tremendous earthquake that topples buildings. A failure caused by residual stresses is often the most difficult to predict and least to be expected.

For mechanical components that operate in severe environments such as a nuclear reactor for an extended time or in safety-critical structures such as an airplane, the presence of residual stresses has a profound influence on the integrity of these components. It is known that one of the main contributing factors for slow-growing cracking in parts exposed to radioactive environments is the presence of residual tensile stresses near the surface. This is a serious issue for containers sealed by welding [77, 46] that contain nuclear wastes, whose radioactive level will remain dangerously high for many centuries. Most shafts or rods are machined by turning, a process that often induces tensile residual stresses near the surface [8], which are detrimental to fatigue life under cyclic loads. On the other hand, the presence of compressive residual stress near the surface is known to enhance fatigue life and inhibit stress-corrosion cracking. For this reason, a process known as shot-peening has been used widely to produce a layer of compressive surface residual stress. However, this process may have some unexpected consequence. Consider a part with a preexisting surface flaw of depth a as shown in Fig. 1.2. After shot-peening, a layer of compressive residual stress is produced in a depth of b, below which the stress becomes tensile. If $b > a$, the flaw under the compressive stress will be fully closed and will not grow unless a substantial external load is applied to open the flaw. The situation changes if $b < a$. The tip of the flaw is now under tensile stress and the compressive stress on the flaw faces up to depth b actually maintains the opening of the flaw. Thus, shot-peening in this case is harmful and the opening of the flaw tip makes it more susceptible to cyclic loading. Furthermore, the detection of the surface flaw by dye penetrant techniques [83] becomes very difficult due to closed flaw faces under compressive residual stress.

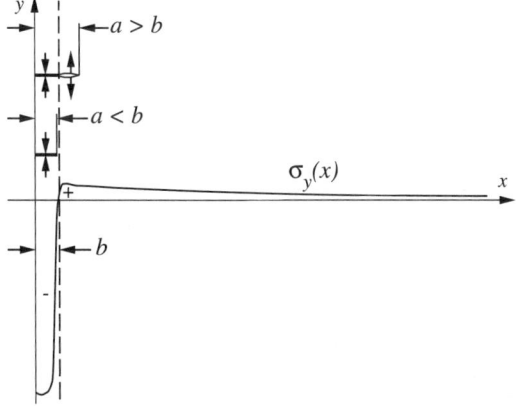

Fig. 1.2. Influence of surface compressive stress on flaws of different sizes.

Residual stress not only affects the initiation and onset of the propagation of surface cracks but also changes the path/growth of a crack as it grows below the surface. First, because the tensile stress near the surface is always balanced by the compressive stress below the surface, the growth of a crack is substantially slowed when it reaches the zone of compressive stress. Second, the growth of the crack often changes the stiffness or compliance of the part and relieves the locked-in load. As a comparison for the first case, a surface crack subjected to uniaxial tension will grow and penetrate a plate when the width of the crack on the surface is about 4 times the thickness [103]. For a crack under the same external load but with a substantial compressive residual stress below the surface, the crack will grow faster along the surface, and the ratio of the width to depth is typically larger than 10. As a consequence, the penetration of the crack is delayed, but the overall integrity of the component keeps weakening as the crack extends further along the surface. When this happens to a pressurized vessel, the occurrence of leakage may signal the onset of a fast moving crack that has already penetrated much of the thickness [53]. For the second case, consider a thin-walled cylinder with a circumferential weld, which introduces a locked-in moment M_o [40]. As a circumferential crack grows, the magnitude of the moment $M(a/t)$ reduces as the compliance of the cracked cylinder increases, as shown in Fig. 1.3. An assumption of constant moment due to residual stress as the crack size increases will certainly lead to an overestimate of the crack growth rate.

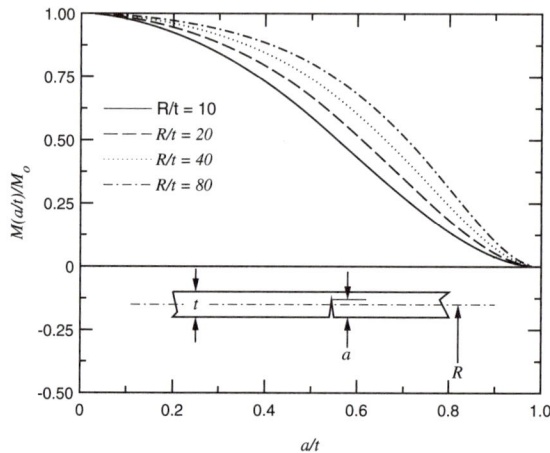

Fig. 1.3. Locked-in moment decreasing with the increase of crack compliance for thin walled cylinders.

Moreover, the distortion of parts after heat-treatment or welding [77] is often so prevalent in manufacturing that it poses as the defining point that separates an experienced engineer from his peers. In the past century our understanding of residual stresses has been greatly enhanced by the analysis based on fundamental mechanics and the measurement of residual stresses that allow us to evaluate and improve the integrity of modern components.

1.3 Mechanical Methods for Residual Stress Measurement

All mechanical methods for residual stress measurement require measuring the deformation due to the release of residual stresses, which are then estimated by using an analysis based on linear elasticity. Therefore, the use of a particular method depends on the availability of not only the means for releasing the stresses and recording the deformation but also the solution and computation for the configuration of the measurement.

Before high-speed computers made numerical solutions for elasticity feasible, analytical solutions were available only for simple geometries, such as plates and cylinders, which led to the development of the boring-out method [64, 82, 113] for cylinders and layer-removal method [122, 126] for beams and plates. In both methods, the stresses to be measured are assumed to vary only in the direction of thickness [14]. As their names imply, the methods release the residual stress by removing material one layer at a time, as shown in Fig. 1.4, and recording the deformation after a layer is removed. The process of repetitive removals of material makes the experimental phase of the methods very time-consuming and prone to errors induced by machining. Once the deformation is obtained, the analysis and computation for estimating the residual stress are fairly straightforward.

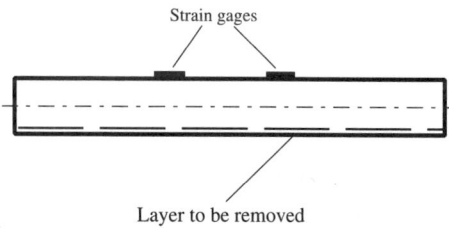

Fig. 1.4. Layer removal method

On the other hand, the hole-drilling method [78, 121, 6] removes very little material. A specially-made high-speed drill [55] makes it a simple procedure to produce a hole with little plastic deformation while the deformation near the hole is recorded. The simplicity in experiment is not without a price. The response to releasing residual stress quickly saturates after the depth of hole reaches about the diameter of the hole, making it only useful for near surface measurement. Nevertheless, the hole-drilling method is the most portable mechanical method that can be used for field applications. Instead of drilling a hole, the trepanning or ring-core method [102] makes use of a circular groove of increasing depth while the deformation at the center of the core is recorded. Because the diameter of the groove is usually much larger than that of a hole, this method allows residual stresses to be measured at a greater depth. The analysis and computation involved in the hole-drilling and trepanning methods range from a simple formula that requires the determination of only two coefficients for measurement of a uniform stress to a sophisticated numerical computation that applies to nonuniform stresses [114].

The sectioning method [128] uses an elaborate procedure to separate a part into sections which are further divided into small pieces. The deformation due to each sectioning is recorded and later used in a three-dimensional (3-D) finite element analysis (FEA) to obtain the residual stresses [129]. Although the method is capable of obtaining a distribution of 3-D residual stresses, the process of repetitive sectioning and installing strain gages makes its implementation very time consuming.

In design or failure analysis, it is often desirable to have the distribution of residual normal stresses obtained on one or more planes of interest. In this case, a more recently-developed technique, the slitting method, is probably the easiest to apply.

The slitting method [19] uses a cut of progressively increasing depth to release the residual stress on a given plane while the deformation is recorded, as shown in Fig. 1.5. It is much faster to carry out than the layer-removal/boring-out and sectioning methods, and applies to both near surface and through-thickness measurements. Table 1.1 shows a comparison of these methods.

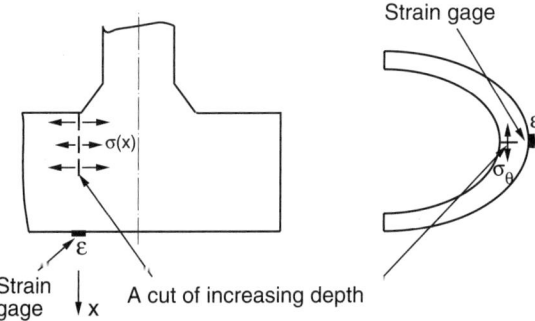

Fig. 1.5. Slitting method measures residual stress distribution on a plane.

Since the beginning of the last century, a considerable effort has been devoted to improving the measurement of residual stresses in beams and cylinders, two of the most-often encountered configurations in industry. The slitting method has greatly simplified the experimental procedure, which can now be carried out in a fraction of the time required by any other method [26, 98].

For a number of applications, the slitting method has found a unique place that no other method can match. They include, for example, the measurements of residual stress in very thin specimens [33, 73], near surface residual stress on a hard-surfaced valve seat where it is unaccessible to a drill, and residual stress on a plane near the toe of a fillet-weld or a welded bracket [22]. In this book we focus our attention on the slitting method. Because the solutions for through-thickness measurement were initially developed using the solutions from linear elastic fracture mechanics (LEFM), the method is also referred to as the crack compliance method. Later in this book, we will use both names interchangeably for through-thickness measurement.

Table 1.1. Comparison of different methods for residual stress measurement

Features	Boring-out Layer removal	Hole-drilling Trepanning	Sectioning	Slitting
Near surface	No[1]	Yes	No	Yes
Through-thickness	Yes	No	Yes	Yes
Experiment	Time-consuming	Simple	Time-consuming	Simple
Computation	Simple	Simple to moderate	Complex	Moderate
In-field application	No	Yes	No	No[2]

[1] Near surface measurement is possible only when combined with the X-ray diffraction method.
[2] Currently no portable cutting machine is available for the slitting method.

1.4 About This Book

Over the past twenty years, the slitting method has been used for a variety of applications, which involve different solutions and computations. The purpose of this book is to present a comprehensive discussion on topics ranging from experimental procedures, analytical solutions, to numerical computations.

Chapter 2 presents the fundamental aspects of the slitting method. General expressions for residual stresses are obtained in Section 2.2 for both a thin cut and a cut of finite width.

Chapters 3-5 cover the basic formulation and computation for compliance functions and are useful if you want to develop your own implementation of the method. The content of the three chapters are independent of each other. So for near-surface measurement one only needs to read Chapter 3, while for through-thickness measurement one only needs to read Chapter 4 or 5.

Chapter 6 introduces different approaches for estimation of the residual stresses. It covers the basic formulation and discusses the limitation associated with each method. The materials presented there apply not only to the slitting method but also to other methods that require stress estimation. For instance, the discussion on the approaches for near surface measurement applies equally to the hole-drilling and trepanning methods.

Chapters 7 and 8 deal with two basic configurations—beams/plates and cylinders. Procedures for strain measurement and stress estimation are presented in detail. The materials in Chapter 7 are also relevant to other configurations of through-thickness measurement, such as the residual normal stress on a plane near the toe of a fillet-weld or a welded bracket.

Chapter 9 introduces the initial strain approach for measurement of residual stresses. The approach represents an alternative to the traditional approximation based on stress. It allows original residual stress to be estimated even if the residual stress in a part to be used in measurement has been partially released.

Chapter 10 discusses the influence of residual stress on fracture strength of glass and surface flaw detection, and the measurement of stress intensity factor variations directly from the strain obtained by the slitting method. The direct measurement of stress intensity factor (SIF) eliminates the need to compute SIF from a measured residual stress distribution.

Appendices A to G provide useful information and procedures for computation of compliance functions.

In addition to those directly relevant to the topics of the book, we have also included in the Bibliography a number of references of general interest to the residual stress measurement; for example, the consistent-splitting method of Rybicki, et al. [112] that extends the layer-removal method to welded cylinders and M. Prime's contour method [99] for measuring the residual axial stress distribution on a cross section of a part in plane strain.

2

Elements of Measurement Using the Slitting Method

2.1 Linear Elasticity and Superposition Principle

All mechanical methods of residual stress measurement are based on the principles of elasticity and linear superposition. In particular, the superposition for the slitting method as shown in Fig. 2.1, is extended from Bueckner's principle for crack propagation [9]. When a cut of depth a is introduced to a part with residual stress (case A), the stress on the site of cut is released (case B). This process is the same as imposing a stress field of the same magnitude of the stress in (case A) with a different sign on the site of the cut (case C), which leads to a stress-free slit face in case B. To compute the deformation or the compliance functions due to introduction of the cut in case B, we make use of case C because there is no change in deformation in case A. Note the superposition shown in Fig. 2.1 remains valid when external loads are present. For a body with prescribed displacement boundary conditions, however, the boundary condition should be properly maintained, as shown in Fig. 2.2. Note that the displacement at the boundary for case C should be set to zero. The stress estimated from the deformation measured from case B and the compliance functions computed from case C is due to both the residual stress and the prescribed boundary condition in case A.

There are situations where the condition of linear elastic deformation may be violated and the superposition is no longer applicable. For example, when a cut of size a is introduced to a thick ring with residual stress $\sigma_o(r)$ from the outer surface shown in Fig. 2.3, the change in stress $\Delta\sigma(r, a)$ near the inner surface 180° from the cut may be considerable and exceed the elastic limit if the increase has the same sign as the original residual stress [45]. In this case, the superposition shown in Fig. 2.1 will no longer be valid because of the presence of the plastic deformation. The superposition principle also assumes that the faces of cut are not in contact during cutting, which may take place if the cut is made into a zone of high compressive stress.

The use of superposition allows residual stress measurements to be carried out on several adjacent planes when the change of stress due to previous cuts

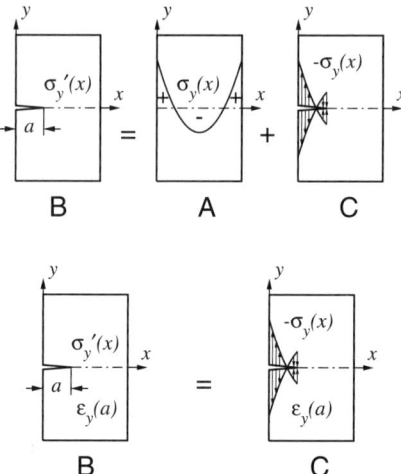

Fig. 2.1. Linear superposition for the slitting method.

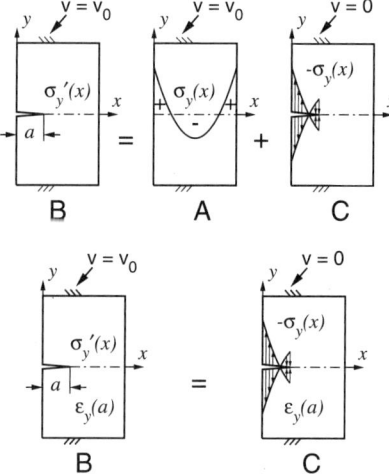

Fig. 2.2. Linear superposition with prescribed displacement conditions for the slitting method.

are included in the final stress estimation. Consider a welded plate, shown in Fig. 2.4, with a residual stress that varies with distance in both thickness and length. To obtain residual normal stress distributions on three planes in the weld and adjacent region, a cut is first made on plane I to measure residual stresses σ_I and τ_I. The release of σ_I and τ_I on plane I changes stresses on planes II and III from σ_{II} and σ_{III} to σ'_{II} and σ'_{III}, which can be measured by using the slitting method on each plane. From superposition, the original

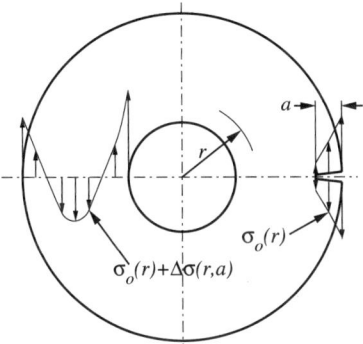

Fig. 2.3. The change of the stress near the inner surface may exceed the elastic limit for a thick-walled ring.

residual stresses on planes II and III are obtained as

$$\sigma_{II} = \sigma'_{II} + \sigma_{II}^{I\sigma} + \sigma_{II}^{I\tau}$$
$$\sigma_{III} = \sigma'_{III} + \sigma_{III}^{I\sigma} + \sigma_{III}^{I\tau} \tag{2.1}$$

where $\sigma_{II}^{I\sigma}$, $\sigma_{II}^{I\tau}$, $\sigma_{III}^{I\sigma}$ and $\sigma_{III}^{I\tau}$ are the stresses computed on planes II and III by applying σ_I and τ_I respectively on the surfaces exposed by cut I. It is important to include residual shear stress τ_I in the computation if the stress field is not symmetric about plane I.

According to Saint-Venant's principle, the influence of releasing a residual stress should be mostly confined in a region of a dimension proportional to the size of the cut. Thus, for through-thickness measurement the change of the stress due to cutting is expected to become very small at a distance about one thickness from the plane of cut. This is confirmed by an analysis presented in Chapter 4 for a beam with an edge-crack.

2.2 Expressions for Approximation of Residual Stresses

Different measurements require different expressions for residual stresses. For a simply-connected 2-D body, shown in Fig. 2.5, an expression for residual normal stresses σ that varies through thickness must satisfy the following conditions

$$\int_0^1 \sigma(x)dx = 0$$

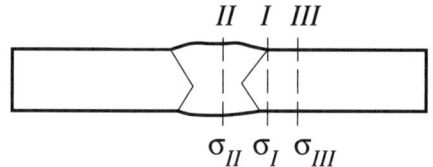

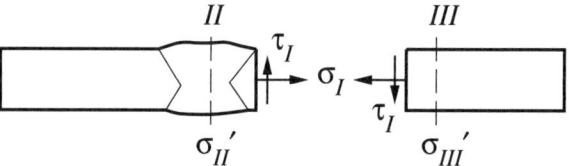

Fig. 2.4. Measurement of residual stresses on several planes in the weld region using the linear superposition.

$$\int_0^1 \sigma(x)(2x-1)dx = 0 \qquad (2.2)$$

and for residual shear stresses τ,

$$\int_0^1 \tau(x)dx = 0$$
$$\tau(0) = \tau(1) = 0 \qquad (2.3)$$

where, for simplicity, the distance x is normalized by the thickness.

It is well known that Legendre polynomials $L_i(x)$ of order $i \geq 2$ always satisfy Eq. (2.2). This can be easily verified by considering the orthogonality which states [70]

$$\int_0^1 L_i(x)L_j(x)dx = \frac{\delta_{ij}}{2i+1} \qquad (2.4)$$

where $\delta_{ij} = 0$ if $i \neq j$. Since $L_0(x) = 1$ and $L_1(x) = 2x - 1$, Eq. (2.2) is guaranteed to hold when $\sigma(x)$ is replaced with $L_i(x)$ with $i \geq 2$. A continuous residual normal stress is thus always expressible by a Legendre polynomial series over the thickness as

$$\sigma(x) = \sum_{i=2}^n A_i L_i(x) \qquad (2.5)$$

where A_i is the amplitude coefficient for $L_i(x)$. In computing Eq. (2.5), the actual expression for a Legendre polynomial is rarely used because it quickly becomes very lengthy. Instead, the recurrence relation is commonly used, which leads to very fast and efficient evaluation of Eq. (2.5). As an example, a subroutine in C programming language is given in Appendix C. It is possible to construct other functions that also satisfy Eq. (2.2) and can be used to represent a continuous residual normal stress, see Section 9.2 for an example. For a multiply-connected 2-D body, such as a ring, one or both of the conditions in Eq. (2.2) may not be required for the residual hoop stress through the wall-thickness.

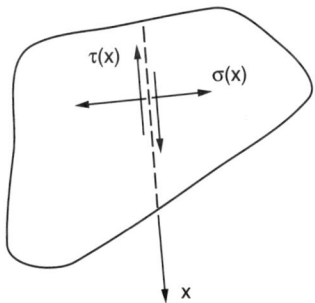

Fig. 2.5. Residual stresses on an arbitrary plane of a free body satisfy the equilibrium conditions.

An expression for residual shear stress, to the authors' knowledge, is not available in literature until recently probably due to much less attention received for the measurement of shear stresses. Its derivation is outlined here. First a general function that satisfies the second condition in Eq. (2.3) may be written as

$$\tau(x) = x(1-x)J(x) \qquad (2.6)$$

which, when substituted into the first condition in Eq. (2.3), gives

$$\int_0^1 \tau(x)dx = \int_0^1 x(1-x)J(x)dx = 0 \qquad (2.7)$$

This is a special case for the orthogonality of a class of Jacobi polynomials [70], which for the nth order is given as

$$J_n(x) = \frac{(-1)^n}{n!x(1-x)}\frac{d^n}{dx^n}\{[x(1-x)]^{1+n}\} = \frac{Q_n(x)}{x(1-x)} \qquad (2.8)$$

Thus, a general residual shear stress may be expressed as

$$\tau(x) = \sum_{i=1}^{n} B_i Q_i(x) \qquad (2.9)$$

where B_i is the amplitude coefficient for the ith order term. Again, the sum can be efficiently evaluated by using the recurrence relation, and a subroutine in C programming language is also included in Appendix C.

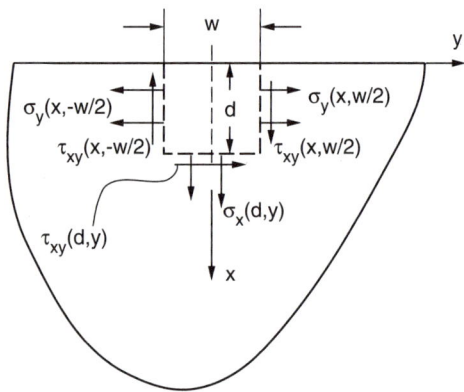

Fig. 2.6. A complete 2-D residual stress field on the site of a cut of finite width.

For near surface measurement no conditions on the resultant force and moment are required. However, the expression that describes the residual stresses released by a cut of finite width is more involved if the cut releases not only normal but other in-plane stresses.

For a slot of width w and depth d as shown in Fig. 2.6, the variation of the normal stress σ_y in a small region that contains the site of the cut may be sufficiently approximated by a second order function in y as below,

$$\sigma_y(x,y) = \sum_{j=0}^{2} y^j f_j(x) \qquad (2.10)$$

where x-axis is chosen to coincide with the centerline of the cut. In practice variables x and y are often normalized by the final depth of cut for a near surface measurement or the thickness of part for through-thickness measurement. It is seen that $f_0(x)$ corresponds to the stress on the centerline of the slit. From equilibrium equations [124] for a 2-D stress field we find

$$\frac{\partial^2 \sigma_x}{\partial x^2} = \frac{\partial^2 \sigma_y}{\partial y^2} \qquad (2.11)$$

which, when combined with Eq. (2.10), leads to

$$\frac{\partial^2 \sigma_x}{\partial x^2} = 2 f_2(x) \tag{2.12}$$

and

$$\sigma_x(x, y) = 2 \int_0^x dx \int_0^x f_2(x) dx + x A(y) + B(y) \tag{2.13}$$

Again, from equilibrium equations

$$\frac{\partial \sigma_x}{\partial x} = -\frac{\partial \tau_{xy}}{\partial y} = 2 \int_0^x f_2(x) dx + A(y)$$

$$\frac{\partial \sigma_y}{\partial y} = -\frac{\partial \tau_{xy}}{\partial x} = f_1(x) + 2y f_2(x) \tag{2.14}$$

After integration we find

$$\tau_{xy}(x, y) = -\int_0^x f_1(x) dx - 2y \int_0^x f_2(x) dx + C(y)$$

$$\tau_{xy}(x, y) = -2y \int_0^x f_2(x) dx - \int A(y) dy + D(x) \tag{2.15}$$

and

$$D(x) = -\int_0^x f_1(x) dx$$

$$C(y) = -\int A(y) dy$$

The shear stress is thus given by

$$\tau_{xy}(x, y) = -2y \int_0^x f_2(x) dx - \int_0^x f_1(x) dx - \int A(y) dy \tag{2.16}$$

At the surface $x = 0$ we have

$$\tau_{xy}(0, y) = 0 \quad and \quad \sigma_x(0, y) = 0 \tag{2.17}$$

Using Eq. (2.17) in Eqs. (2.13) and (2.16) yields

$$B(y) = 0 \quad and \quad \int A(y) dy = 0 \tag{2.18}$$

Thus, we may set $A(y) = 0$. The final form of the stresses becomes

$$\sigma_x(x, y) = 2 \int_0^x dx \int_0^x f_2(x) dx$$

$$\sigma_y(x, y) = f_0(x) + y f_1(x) + y^2 f_2(x)$$

$$\tau_{xy}(x, y) = -2y \int_0^x f_2(x) dx - \int_0^x f_1(x) dx \tag{2.19}$$

Equation (2.19) shows the relationship among the three stress components in the region of the cut. For a very thin cut, $w/d \approx 0$, we have

$$\sigma_x(x,0) = f_0(x)$$
$$\tau_{xy}(x,0) = -\int_0^x f_1(x)dx \tag{2.20}$$

where σ_x is omitted because it only acts on the bottom of cut. For a cut of finite width, the stresses to be released on the side faces of the cut $(y = \pm w/2)$ are given by

$$\sigma_y(x, \pm\frac{w}{2}) = f_0(x) \pm \frac{w}{2} f_1(x) + \frac{w^2}{4} f_2(x)$$
$$\tau_{xy}(x, \pm\frac{w}{2}) = \mp w \int_0^x f_2(x)dx - \int_0^x f_1(x)dx \tag{2.21}$$

and on the bottom of the cut of depth $x = d$ by

$$\sigma_x(d,y) = 2 \int_0^d dx \int_0^x f_2(x)dx$$
$$\tau_{xy}(d,y) = -2y \int_0^d f_2(x)dx - \int_0^d f_1(x)dx \tag{2.22}$$

Equation (2.21) shows that, as width w increases, the normal and shear stresses acting on one side of cut may become different from those on the other side. The expressions for σ_x and τ_{xy} on the bottom of cut in Eq. (2.22) represent a zero-order approximation and a linear approximation respectively. Note that stresses estimated using Eqs. (2.21) and (2.22) automatically satisfy the equilibrium conditions along the faces of the cut. Although Figure 2.3 shows a cut with a flat bottom, the solutions obtained above are equally valid for a cut with a curved bottom.

Equations (2.21) and (2.22) are also useful for through-thickness measurement when other stress components are not negligibly small. For a thin cut in particular, the zero-resultant force conditions over the thickness are satisfied when Eqs. (2.5) and (2.9) are combined with Eq. (2.20). That is,

$$\sigma_x(x,0) = f_0(x) = \sum_{i=2}^{n} A_i L_i(x)$$
$$\tau_{xy}(x,0) = -\int_0^x f_1(x)dx = \sum_{i=1}^{n} B_i Q_i(x) \tag{2.23}$$

the second of which leads to

$$f_1(x) = -\sum_{i=1}^{n} B_i \frac{dQ_i(x)}{dx} \tag{2.24}$$

So far we have limited our discussion solely to 2-D stresses. For three-dimensional (3-D) stresses, a general expression for residual stresses that satisfies all the equilibrium conditions over a cross-section of an arbitrary shape is not yet available. Fortunately, the initial strain approach to be introduced in Chapter 9 provides a useful alternative to the conventional stress-based approach and allows a rigorous description of the residual stresses in parts of complex geometries.

After we have constructed an expression to approximate the residual stresses to be measured, the strain response to each function defined in the expression with a unit magnitude can be computed using one of the approaches to be presented in Chapters 3, 4 and 5. The strain obtained as a function of the depth of cut is referred to as the compliance function, and, therefore, we sometimes refer to the slitting method as the compliance method.

2.3 Experimental Procedures

Deformation due to releasing residual stress by a cut of progressively increasing depth can be measured as the change of displacements and/or strain. The latter one is by far the most commonly measured variable thanks to the wide availability of high precision electric-resistance strain gages of various sizes and patterns [41]. Although measurement of strain using strain gages offers a higher sensitivity and reliability than most displacement based measurements, it has certain limitations:

1. Measurement is limited to a few fixed locations;
2. As the number strain gages increases, the soldering and cabling of the gages becomes tedious and time-consuming;
3. Sensitive to temperature change if the gage's thermal expansion coefficient does not match that of the specimen adequately.

Fortunately, the slitting method in most cases requires measurement of strain only at one or two locations, as shown in Fig. 2.7A. When choosing a strain gage, it is crucial to match the thermal expansion coefficient of the strain gage with that of the surface on which the gage will be installed. Also, the gage length needs to be short enough to reduce the influence of strain gradient and increase the sensitivity of the measurement.

The method used to make a cut of increasing depth has evolved from sawing [71, 72], milling [11, 15, 16] to electric discharge machining (EDM) [38]. Sawing and milling are universally available but the cutting may introduce unwanted temperature increase and plastic deformation near the bottom of the cut. To reduce the effect of clamping force on the measurement, the plane of cut should be located sufficiently away from the fixture. A precise measurement of the depth of cut is more difficult when sawing is used. Furthermore, the release of a compressive residual stress may produce significant "binding" on the cutter to cause breakage, which often terminates a measurement prematurely. Electric discharge machining makes a cut without applying any forces, which minimizes the clamping force required to secure the specimen. For wire EDM, the location and depth of cut can be controlled precisely and cutting can be resumed in most cases after rethreading the wire if it breaks. For near surface measurement on a curved surface, conventional EDM is uniquely qualified to make a cut of nearly uniform depth, as shown in Fig. 2.7B, using a sheet of electrode that has a profile matching the curved surface. Because the electrode wears out gradually during cutting, the measurement of cut depth needs to be calibrated carefully for a given material and a given set of cutting conditions. The use of EDM has two main limitations:

1. It only cuts electrical conductive materials;
2. It is not portal for field applications.

There are two situations that require special attention during measurements. When a thin cut is made by a wire through a region of high compressive stress, the deformation due to releasing the stress may be so large that the faces of cut become in contact, which invalidates the assumption of the superposition principle. This situation can be corrected easily by cutting backward to remove the material in contact. On the other hand, a thin cut in a region of high tensile stress may initiate crack propagation near the tip of the cut, which terminates the test prematurely. In spite of these limitations, EDM remains the best method for making a high precision cut of progressively increasing depth for electrical conductive materials. For other materials, however, a mechanical method of cutting has to be considered.

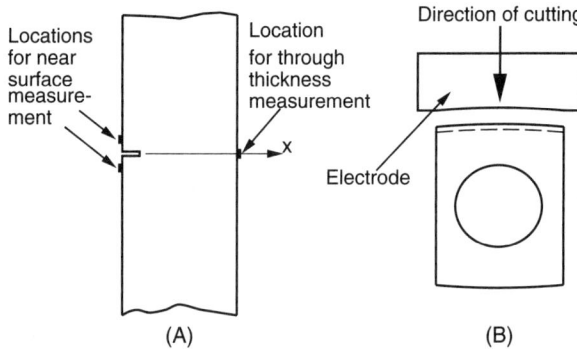

Fig. 2.7. (A) Locations for strain measurement. (B) Use of EDM to make a cut on a curved surface for near surface stress measurement.

3

Compliance Functions for Near-Surface Measurement: The Body Force Approach

3.1 Introduction

When a crack or a cut of finite width is introduced to a surface, the release of the residual stresses on the plane of the cut leads to deformation that can be measured and used to predict the residual stress that existed on the plane before the cut was made. Since a cut of finite width is much easier to introduce than a crack, and less likely to experience face closure associated with releasing compressive stresses, it is of practical importance to obtain the compliance functions for a cut of finite width. For this reason, the approach based on a cut of finite width introduced in this chapter for near surface stress measurement is often referred to as the slitting method.

In this chapter the strain on the free surface of a semi-infinite plane due to a notch or a cut of finite width will be obtained by using the body force method developed by Nisitani [86]. The body force method has been used to obtain the local stress field near a notch for various geometries [87]. The method makes use of stress solutions for point forces acting in an unnotched body. The surface traction on the faces of the notch is taken to be the same as that in the unnotched body but with an opposite sign. The rest of the boundary of the notched body is taken as traction-free which is always satisfied by the point force solutions. It is therefore particularly convenient to apply the body force method to a body with only residual stresses. Unlike the earlier work by Nisitani which focused on the local stress field near the notch tip, the present study aims at obtaining the strain along the free surface due to surface traction on the faces of a cut.

Although the numerical formulation for the body force method is similar to that for the boundary element (boundary integral) method, the two methods were developed independently. The boundary element method usually uses fundamental solutions for displacements while the body force method uses solutions for stresses. It is thus more straightforward to use the body force method to solve the problem for a slot subjected to loading on its faces.

3.2 Analysis

Consider a semi-infinite plane with a cut as shown in Fig. 3.1. The surface traction on the faces of the cut corresponds to the residual stresses released by making the cut. To achieve the same boundary conditions in a semi-infinite plane shown in Fig. 3.2, we introduce distributed forces $f_x(s,t)$ and $f_y(s,t)$ in the x and y directions to the contour (s,t) of the cut and adjust the distributions of f_x and f_y such that the resulting stresses at the site of the cut coincide with the surface traction shown in Fig. 3.1. For a cut of width $2w$ and depth d the normal and shear stresses σ and τ on the sides of the cut can be expressed as

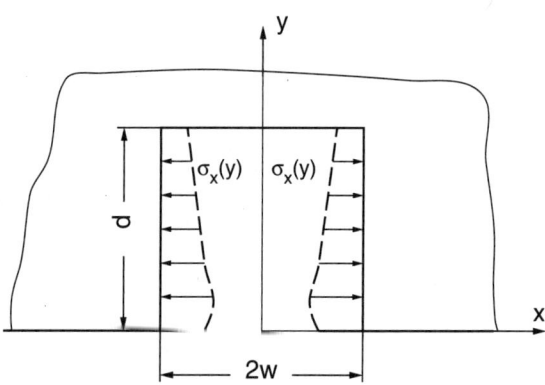

Fig. 3.1. A cut of finite width introduced into a semi-infinite plane with loading on the faces of the cut.

$$
\begin{aligned}
\tau_{xy}(\pm w, y) = & \int_{-w}^{w} [f_x(s,d)H_x(\pm w, y, s, d) \\
& +f_y(s,d)H_y(\pm w, y, s, d)]ds \\
& + \int_{0}^{d} [f_x(w,t)H_x(\pm w, y, w, t) \\
& +f_y(w,t)H_y(\pm w, y, w, t)]dt \\
& + \int_{0}^{d} [f_x(-w,t)H_x(\pm w, y, -w, t) \\
& +f_y(-w,t)H_y(\pm w, y, -w, t)]dt
\end{aligned}
\tag{3.1}
$$

$$\sigma_x(\pm w, y) = \int_{-w}^{w} [f_x(s,d)F_x(\pm w, y, s, d)$$
$$+ f_y(s,d)F_y(\pm w, y, s, d)]ds$$
$$+ \int_{0}^{d} [f_x(w,t)F_x(\pm w, y, w, t)$$
$$+ f_y(w,t)F_y(\pm w, y, w, t)]dt$$
$$+ \int_{0}^{d} [f_x(-w,t)F_x(\pm w, y, -w, t)$$
$$+ f_y(-w,t)F_y(\pm w, y, -w, t)]dt \qquad (3.2)$$

and the stresses on the bottom of the cut can be expressed as

$$\sigma_y(x, d) = \int_{-w}^{w} [f_x(s,d)G_x(x, d, s, d)$$
$$+ f_y(s,d)G_y(x, d, s, d)]ds$$
$$+ \int_{0}^{d} [f_x(w,t)G_x(x, d, w, t)$$
$$+ f_y(w,t)G_y(x, d, w, t)]dt$$
$$+ \int_{0}^{d} [f_x(-w,t)G_x(x, d, -w, t)$$
$$+ f_y(-w,t)G_y(x, d, -w, t)]dt \qquad (3.3)$$

$$\tau_{xy}(x, d) = \int_{-w}^{w} [f_x(s,d)H_x(x, d, s, d)$$
$$+ f_y(s,d)H_y(x, d, s, d)]ds$$
$$+ \int_{0}^{d} [f_x(w,t)H_x(x, d, w, t)$$
$$+ f_y(w,t)H_y(x, d, w, t)]dt$$
$$+ \int_{0}^{d} [f_x(-w,t)H_x(x, d, -w, t)$$
$$+ f_y(-w,t)H_y(x, d, -w, t)]dt \qquad (3.4)$$

in which $F_x(x, y, s, t)$, $F_y(x, y, s, t)$, $G_x(x, y, s, t)$, $G_y(x, y, s, t)$, $H_x(x, y, s, t)$ and $H_y(x, y, s, t)$ are point force solutions and have been obtained by Nisitani [87]. For the configuration shown in Fig. 3.2 they may be simplified as

$$F_x = \frac{-1}{4\pi y} [\frac{l(3l^2 + m^2)}{(l^2 + m^2)^2}$$
$$+ \frac{5l^5 + 4l^3(n^2 - n + 1) - ln^2(n^2 - 12n + 12)}{(l^2 + n^2)^3}]$$

$$F_y = \frac{-1}{4\pi y}[\frac{m(l^2 - m^2)}{(l^2 + m^2)^2}$$
$$+ \frac{l^4(n+6) + 4l^2 n(2n^2 + 3n - 3) + n^3(7n^2 - 10n + 4)}{(l^2 + n^2)^3}]$$

$$G_x = \frac{1}{4\pi y}[\frac{l(l^2 - m^2)}{(l^2 + m^2)^2}$$
$$- \frac{l^5 + 4l^3(3n - 1) - ln^2(n^2 + 4n - 12)}{o}ver(l^2 + n^2)^3]$$

$$G_y = \frac{-1}{4\pi y}[\frac{m(l^2 + 3m^2)}{(l^2 + m^2)^2}$$
$$+ \frac{l^4(n-2) + 4l^2 n(n^2 - 3n + 3) + n^3(3n^2 + 6n - 4)}{(l^2 + n^2)^3}]$$

$$H_y = \frac{-1}{4\pi y}[\frac{l(l^2 + 3m^2)}{(l^2 + m^2)^2}$$
$$- \frac{l^5 + 4l^3(n^2 - n - 1) + ln^2(3n^2 - 20n + 12)}{(l^2 + n^2)^3}]$$

$$H_x = \frac{-1}{4\pi y}[\frac{m(3l^2 + m^2)}{(l^2 + m^2)^2}$$
$$+ \frac{l^4(3n + 2) + 4l^2 n(n^2 - 3n + 3) + n^3(n^2 + 2n - 4)}{(l^2 + n^2)^3}]$$

in which

$$l = \frac{x - s}{y}, \quad m = \frac{y - t}{y}, \quad n = \frac{y + t}{y}$$

Generally, the unknown functions $f_x(s,t)$ and $f_y(s,t)$ which simultaneously satisfy Eqs. (3.1) to (3.4) have to be solved numerically. Because of the symmetry about the y-axis we divide the depth d and half width w into $n = n_1 + n_2$ intervals and approximate the functions f_x and f_y in each interval with a uniform stress. Denoting the magnitudes of f_x and f_y in the ith interval as f_{xi} and f_{yi}, Equations (3.1) to (3.4) then become

$$\sigma_x(\pm w, y) = \sum_{i=1}^{n_2} \int_{s_{i-1}}^{s_i} [f_{xi} F_x(\pm w, y, s, d) + f_{yi} F_y(\pm w, y, s, d)]ds$$
$$+ \sum_{i=1}^{n_1} \left\{ \int_{t_{i-1}}^{t_i} [f_{xi} F_x(\pm w, y, w, t) + f_{yi} F_y(\pm w, y, w, t)]dt \right.$$
$$\left. + \int_{t_{i-1}}^{t_i} [f_{xi} F_x(\pm w, y, -w, t) + f_{yi} F_y(\pm w, y, -w, t)]dt \right\} \quad (3.5)$$

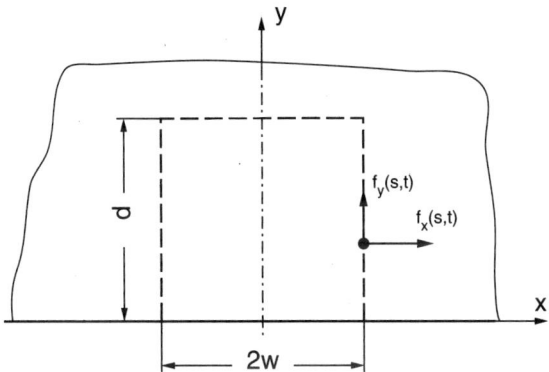

Fig. 3.2. A pair of line forces applied on location (s, t) on the faces of the cut.

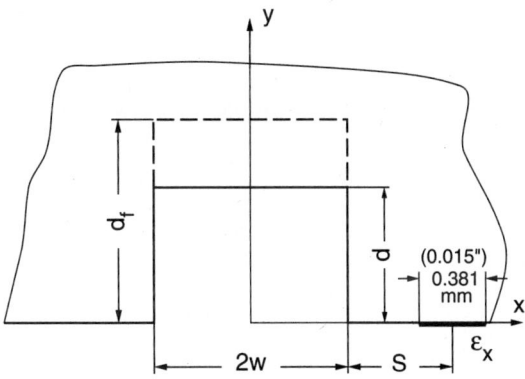

Fig. 3.3. A strain gage located near the cut for measuring strain changes due to cutting.

$$\tau_{xy}(\pm w, y) = \sum_{i=1}^{n_2} \int_{s_{i-1}}^{s_i} [f_{xi} H_x(\pm w, y, s, d) + f_{yi} H_y(\pm w, y, s, d)] ds$$

$$+ \sum_{i=1}^{n_1} \left\{ \int_{t_{i-1}}^{t_i} [f_{xi} H_x(\pm w, y, w, t) + f_{yi} H_y(\pm w, y, w, t)] dt \right.$$

$$+ \int_{t_{i-1}}^{t_i} [f_{xi}H_x(\pm w, y, -w, t) + f_{yi}H_y(\pm w, y, -w, t)]dt \Bigg\} \quad (3.6)$$

$$\sigma_y(x, d) = \sum_{i=1}^{n_2} \int_{s_{i-1}}^{s_i} [f_{xi}G_x(x, d, s, d) + f_{yi}G_y(x, d, s, d)]ds$$

$$+ \sum_{i=1}^{n_1} \Bigg\{ \int_{t_{i-1}}^{t_i} [f_{xi}G_x(x, d, w, t) + f_{yi}G_y(x, d, w, t)]dt$$

$$+ \int_{t_{i-1}}^{t_i} [f_{xi}G_x(x, d, -w, t) + f_{yi}G_y(x, d, -w, t)]dt \Bigg\} \quad (3.7)$$

$$\tau_{xy}(x, d) = \sum_{i=1}^{n_2} \int_{s_{i-1}}^{s_i} [f_{xi}H_x(x, d, s, d) + f_{yi}H_y(x, d, s, d)]ds$$

$$+ \sum_{i=1}^{n_1} \Bigg\{ \int_{t_{i-1}}^{t_i} [f_{xi}H_x(x, d, w, t) + f_{yi}H_y(x, d, w, t)]dt$$

$$+ \int_{t_{i-1}}^{t_i} [f_{xi}H_x(x, d, -w, t) + f_{yi}H_y(x, d, -w, t)]dt \Bigg\} \quad (3.8)$$

We now specify the values of the stresses at the left hand side of Eqs. (3.5) to (3.8) as the mean values of the n intervals. This leads to $2n = 2(n_1 + n_2)$ linear equations which can be solved for the unknown coefficients f_{x1} to f_{xn} and f_{y1} to f_{yn}. The stress distribution at the surface is then given by

$$\sigma_x(x, 0) = \sum_{i=1}^{n_2} \int_{s_{i-1}}^{s_i} [f_{xi}F_x(x, 0, s, d) + f_{yi}F_y(x, 0, s, d)]ds$$

$$+ \sum_{i=1}^{n_1} \Bigg\{ \int_{t_{i-1}}^{t_i} [f_{xi}F_x(x, 0, w, t) + f_{yi}F_y(x, 0, w, t)]dt$$

$$+ \int_{t_{i-1}}^{t_i} [f_{xi}F_x(x, 0, -w, t) + f_{yi}F_y(x, 0, -w, t)]dt \Bigg\} \quad (3.9)$$

Since the integrals given in Eqs. (3.5) to (3.9) can be carried out in a closed form, all the terms on the right hand side of these equations can be evaluated quickly and accurately.

The strain distribution $\epsilon_x(x, 0)$ along the free surface can be directly calculated from Hook's law, i.e.,

$$\epsilon_x(x, 0) = \sigma_x(x, 0)/E' \quad (3.10)$$

where $E' = E$ for plane stress and $E' = E/(1 - \nu^2)$ for plane strain with E the elastic modulus and ν the Poisson's ratio.

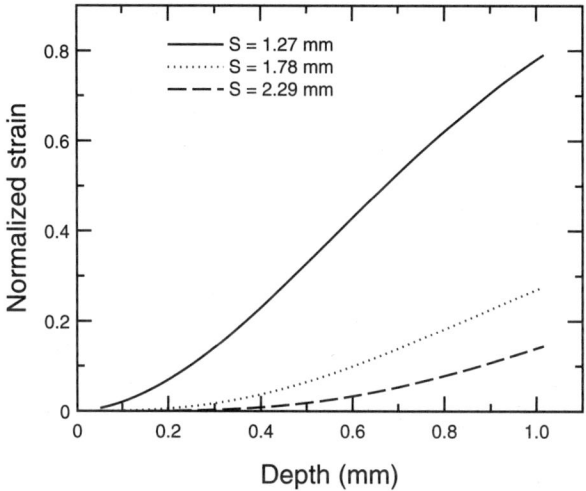

Fig. 3.4. Strain measured at different locations for a uniform stress.

3.3 Results

The body force method has been shown by Nisitani [87] to be able to obtain accurate results for notch problems when the number of discretization n is larger than 48. For the present configuration the values of calculated strain were found to converge rapidly when n is larger than 64. Thus, a number of 192 divisions for one half of the slot is used for the subsequent computation which can be carried out on a personal computer.

The surface traction shown in Fig. 3.1 which is due to external loading or residual stresses may be expressed in terms of an Nth order power series or an orthogonal polynomial series, i.e.,

$$\sigma_x(y) = \sum_{i=0}^{N} \sigma_i P_i(\frac{y}{d_f}) \tag{3.11}$$

where d_f is the final depth of the cut and σ_i is the amplitude of the stress for the ith term P_i. Substituting Eq. (3.11) into Eq. (3.5) and setting the left hand side of the Eqs. (3.6), (3.7) and (3.8) to zero, the $2n$ unknowns for the loading condition shown in Fig. 3.1 can be solved. For convenience of the comparison between the results for a slot and a crack, we will use power functions for $P_i(y/d_f)$ in Eq. (3.11). However, as will be shown later in Chapter 6, orthogonal polynomials such as Legendre polynomials are superior to the power functions for residual stress measurement.

We now obtain the strain ϵ_x on the free surface at a distance S from the side of the cut using Eqs. (3.9) and (3.10) with E' equal to unity. To simulate the strains measured in a typical experiment, see Appendix F, the average

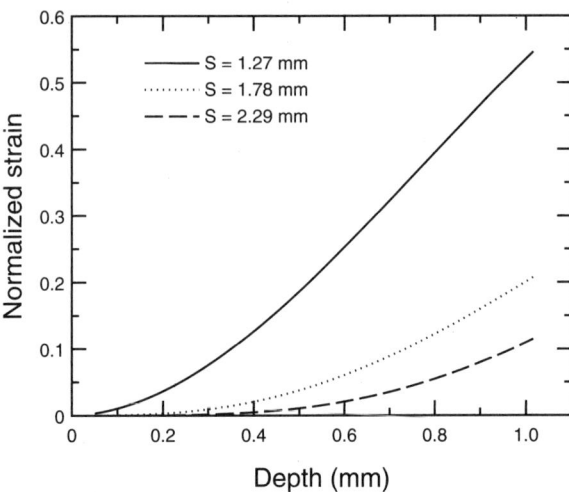

Fig. 3.5. Strain measured at different locations for a linearly varying stress.

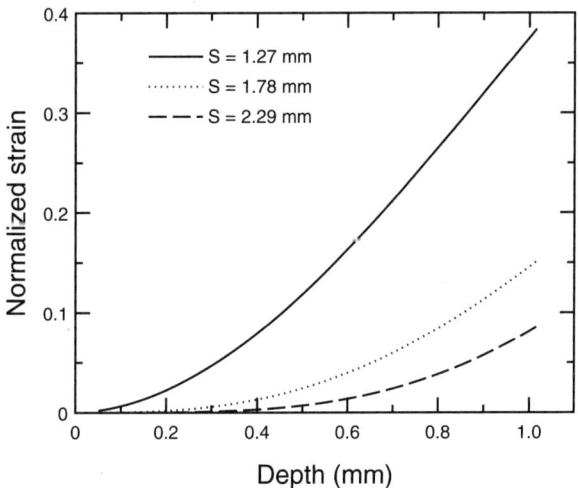

Fig. 3.6. Strain measured at different locations for a quadratically varying stress.

strains of ϵ_x over a strain gage of length 0.381 mm, as shown in Fig. 3.3, are calculated for distance $S = 1.27$ mm, 1.778 mm, 2.286 mm and 2.794 mm. The width of the cut is taken as 0.1016 mm and the depth from 0.0508 mm to 1.016 mm. Figures 3.4 to 3.6 show the calculated strains for uniform, linear and quadratic stress distributions with σ_0 to σ_2 being unity.

The deformation due to an edge crack in a semi-infinite plane has been obtained by Cheng and Finnie [34]. Figure 3.7 shows the ratios of the strains for a cut of finite width to that for a crack. It is seen that, as the ratio of depth to width increases, the strain response to a uniform stress for a cut reduces to that for a crack faster than that to non-uniform stresses. Also, when the ratio of depth to width is less than five, the difference between a cut and a crack increases rapidly, and the use of crack compliance functions in this case may lead to a significant error.

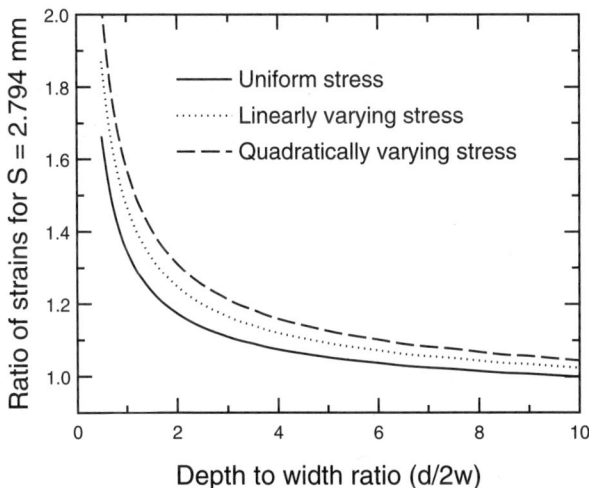

Fig. 3.7. Ratio of the strain response for a slot to that for a crack subjected to different stresses.

We now turn to the hole-drilling method which has widely been used for near surface residual stress measurement and has been extended to measuring stresses varying with depth [114]. Since there is no analytical solution available for a blind hole, the relationship between measured strains and released stresses has to be either calibrated experimentally [6, 107] or calculated numerically by the finite element method [114, 7, 120, 4, 55, 115]. To compare the sensitivity of the present method with that of the hole-drilling method, two stress states, uniaxial stress and bi-axial stress with equal amplitude, are considered here.

For the hole-drilling method the strain is assumed to be taken from a strain gage which coincides with the direction of the maximum principal stress. The maximum hole radius and depth allowed for a standard strain gage rosette configuration [81] is used in the calculation, i.e. $r/r_m = 0.5$ and $h/r_m = 0.5$ where h is depth of the hole, r and r_m are the hole radius and the mean radius of the strain gage rosette, respectively. This configuration gives the maximum strain response. For a smaller ratio of r/r_m with the same hole depth, the strain response may reduce considerably from that calculated based on $r/r_m = 0.5$. The maximum ratio of r/r_m specified in the ASTM Standard

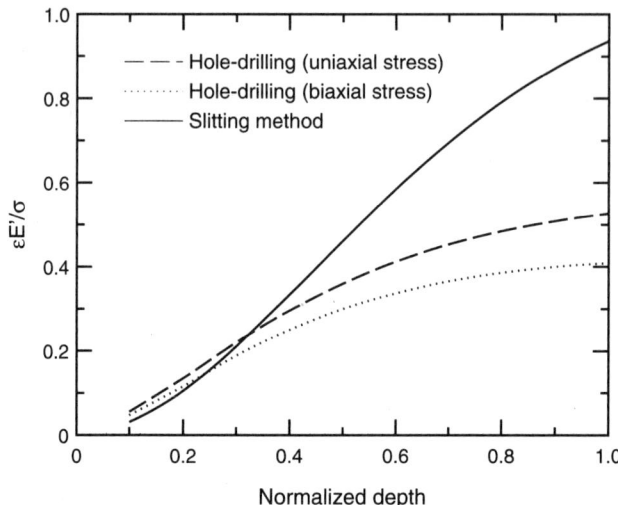

Fig. 3.8. Comparison of the normalized strain responses for the hole drilling and slitting methods for a uniform stress.

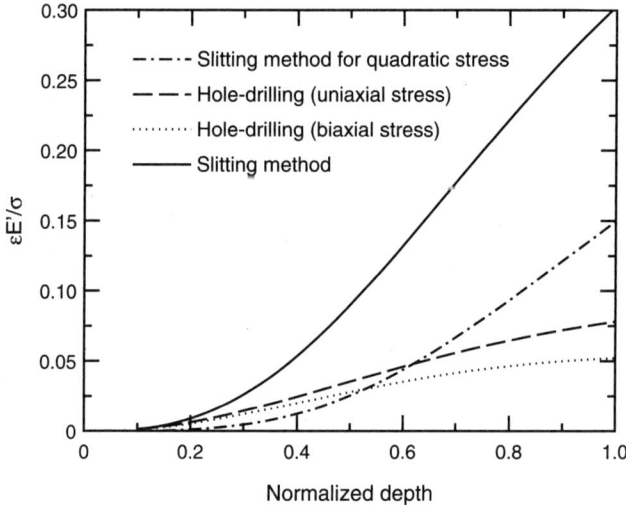

Fig. 3.9. Comparison of the normalized strain responses for the hole drilling and slitting methods for a linearly varying stress.

[134] is 0.4 which leads to a strain response about 44% less than that based on the maximum hole radius.

For the slitting method the average strain determined in the region $S/d = 1$ for a cut of width $w = 0.1d$ is used. The direction of the cut is assumed to be perpendicular to the direction of the maximum principal stress. For simplicity the magnitude of the stresses, Young's modulus and depth of the cut or the hole are taken as unity and the Poisson's ratio $\nu = 0.3$. The numerical values given in [114] are used for an estimate of strains for the hole-drilling method. The estimated strain response for a uniform stress field is shown in Fig. 3.8 as a function of the hole depth or the cut depth. For a thin cut the strain response for the slitting method depends only on the normal stress released along the cut. Thus, the same response is obtained for both stress states. For the hole-drilling method the response depends on the combination of the principal stresses. Thus, two responses are obtained for the two stress states. In Fig. 3.9 the strain responses to a linear and a quadratic stress distribution are shown for the slitting method. Also shown in Fig. 3.9 is the strain response to the linear stress distribution for the hole-drilling method.

It is seen from Figs. 3.8 and 3.9 that the strain response for the slitting method is about 44% higher for the uniform stress and 300% higher for the linear stress than that for the hole-drilling method. Even for the quadratic stress the crack compliance method leads to a response larger than that of the hole-drilling method for the linear stress. This indicates that the present method is more sensitive and capable of estimating a higher order stress field than the hole-drilling method. The other major difference between the two methods is that the response of the hole-drilling method to the release of residual stresses decreases rapidly as the hole depth increases while the response of the present method decreases only slightly. Thus, an improved estimation of residual stresses as a function of the depth below the surface should be achieved by using the slitting method.

3.4 Comparison of Cuts with Circular Bottom and Flat Bottom

The most commonly used process to introduce a cut of finite width is wire electric discharge machining (wire EDM) [38], which leads to a cut with a semicircular bottom. However, to simplify the analysis, the numerical model for the cut shown in Fig. 3.1 is based on a slot with a flat bottom. To improve the approximation, the width of the slot used in computation is reduced from $2w$ to $2w_e$ so that the area becomes the same as that for the cut with semicircular bottom. That is,

$$\frac{w_e}{w} = \frac{w}{2d}\left\{\cos^{-1}(1-\frac{d}{w}) - (1-\frac{d}{w})\sin[\cos^{-1}(1-\frac{d}{w})]\right\} \quad d < w$$

$$\frac{w_e}{w} = \frac{1-w}{d}(1-\frac{\pi}{4}) \qquad d \geq w \qquad\qquad (3.12)$$

It is still necessary to examine the influence of the discrepancy between the two geometries on compliance computation. As an approximation, the semicircular bottom is divided into a number of small intervals with boundary being defined by short straight lines, as shown in Fig. 3.10. As the number of intervals increases the boundary of the intervals reduces to a semicircular shape. Using this approach, the compliance of a cut with a semicircular bottom may be obtained using the body force method [86]. As a comparison, the compliance, i.e., the strains normalized by σ/E' for a uniform stress are calculated for both cases. $E' = E/(1 - \nu^2)$ is the plane strain elastic modulus. The width of the cut with right corner bottom is determined by Eq. (3.12). Figure 3.11 shows the normalized strains as a function of depth normalized by one-half width w. It is seen the difference becomes very small when d/w is larger than 2. Also, the maximum difference at $d/w = 1$ is found to be only about 7 percent. This error is, therefore, not expected to affect the stress estimation using a least squares fit, which is described in detail in Chapter 6.

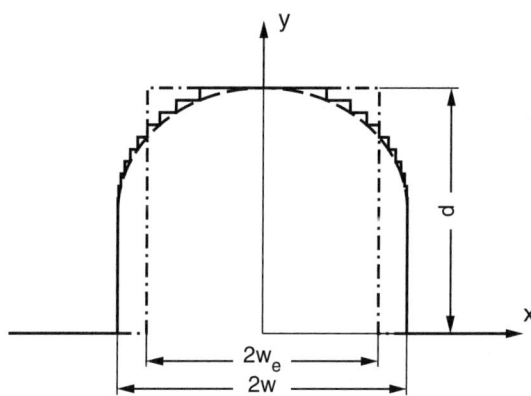

Fig. 3.10. A semi-circular bottom approximated by a series of steps of decreasing sizes.

3.5 Discussion

The approach presented in this chapter for computing the compliances of a cut of finite width is based on the body force method (BFM). Later in Chapter 5, we will present an alternative approach based on the finite element method (FEM) for measuring residual stresses through-the-thickness. When compared with FEM, the advantages of the BFM are:

1. Use of analytical solutions leads to more accurate computation for a slit in an semi-infinite plane, which can be carried out more rapidly on a personal computer.

2. Loading conditions on the faces of the slit can be easily implemented in terms of either a continuous function or piecewise function.

3. Modelling of a cut of progressively increasing depth can be performed directly for any given values of depths.

4. The strain on the free surface can be computed at an arbitrary location from the cut.

The limitations of BFM are:

1. For geometries other than a semi-infinite plane the analysis becomes very complicated to implement.

2. The method is currently limited to parts of isotropic material properties or those with a single surface layer [95] for which the material properties differ from the rest of the part.

3. Few computer programs based on BFM are commercially available.

4. Not suitable for approximations using initial strain fields as described in Chapter 9.

Since the assumption of a semi-infinite plane with isotropic material properties is valid for the majority of applications of near surface residual stress measurement, BFM has been used extensively for this purpose.

Apart from presenting the method for computing the compliance functions, we also demonstrated in this chapter that for measuring nonuniform near-surface stresses the slit compliance method is more sensitive than the hole-drilling method. This feature is especially useful for measuring residual stresses of rapidly varying distribution. It also allows residual stresses to be measured at a greater depth than is possible for hole- drilling. Moreover, unlike the hole-drilling method for which misalignment of the hole with respect to the strain gage rosette may lead to a significant error in stress estimation, the slitting method uses the actual distance measured from the cut to the strain gage to compute compliance functions and estimate the residual stresses. This is possible because the compliance functions can be determined directly at an arbitrary location on the free surface.

Additionally, the compliance functions for the measurement of shear stresses can also be obtained using the same approach presented here. In this case another strain gages will be installed on the other side of the cut shown in Fig. 3.3 to allow strain variations measured on both sides of the cut. If the stress is nonuniform along the length of the cut, strain variations in both depth and length can be measured using an array of gages installed along the cut, as shown in Fig. 3.12. Then the average stress in the vicinity of each pair of strain gages can be estimated using the 2-D compliance functions.

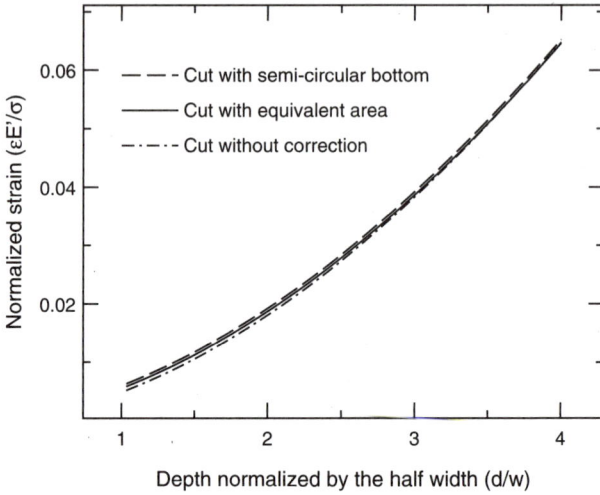

Fig. 3.11. Comparison of the normalized strain for three configurations.

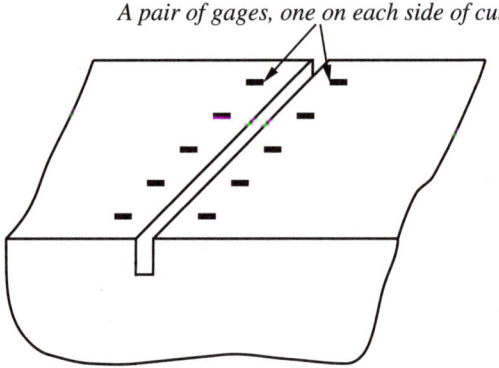

Fig. 3.12. An array of gages used to measure near-surface stress variation in directions of both depth and length of the cut.

4

Compliance Functions for Through-Thickness Measurement: The LEFM Approach

4.1 Introduction

Crack compliances, as discussed throughout this book, are referred to as the unit elastic response of a cracked body to surface tractions acting on the crack faces. The surface tractions may be those actually applied on the crack faces or the residual stresses released by introducing a thin cut. Two approaches have been used to obtain crack compliances as a function of crack sizes for an arbitrary surface traction. One uses the stress intensity factor solutions from linear elastic fracture mechanics (LEFM). Another uses numerical computations based on the finite element method, which will be introduced in the next chapter.

In this chapter we will demonstrate how to obtain crack compliance functions for an edge-cracked beam and an edge-cracked disk using Castigliano's theorem [62] and virtual work from stress intensity factor solutions. Then the concept of a virtual cracked-element is used to obtain the solution for a complete circumferential crack in a long cylinder and for a radial crack in a ring or a cylinder.

4.2 An Edge-Cracked Beam

4.2.1 Normal Stress on Crack Faces

Consider a beam with an edge crack of size a, which is subjected to a normal surface traction $\sigma_y(x)$ as shown in Fig. 4.1. For simplicity in the discussion that follows we take the thickness of the beam as unity, i.e. $t = 1$. Thus, for a problem with $t \neq 1$, the dimensions shown in Fig. 4.1 correspond to those normalized by its thickness. We will use superscripts u and l to denote displacements on the upper and the lower edges respectively. At a distance s from the edge crack the displacement can be decomposed into two components: u in the x-direction and v in the y-direction. To obtain the horizontal

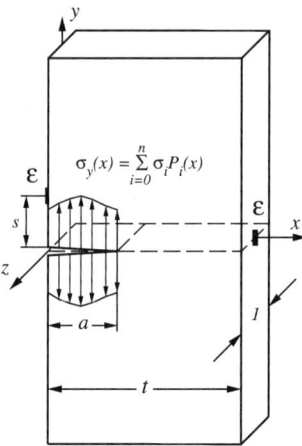

Fig. 4.1. Strains at locations near the cut or on the back face of the beam due to surface traction on the faces of the edge-crack.

displacement v^u at the upper edge, a pair of virtual horizontal line forces[1] F and a pair of virtual horizontal line forces Q of per unit length are introduced at the locations where the displacements are to be determined as shown in Fig. 4.2. The change of the strain energy U [123] due to the presence of the crack is given by

$$U(a, s) = \frac{t}{2E'} \int_0^a [K_I(a) + K_I^f(a, s) + K_I^q(a, s)]^2 da \qquad (4.1)$$

in which $E' = E$ for plane stress and $E' = E/(1 - \nu^2)$ for plane strain with ν being the Poisson's ratio. $K_I(a)$, $K_I^f(a, s)$ and $K_I^q(a, s)$ are the mode I stress intensity factors due to the normal stress $\sigma_y(x)$ and virtual forces F and Q respectively.

Using Castigliano's theorem and setting virtual forces F to zero leads to the expression for the horizontal displacement at the upper edge

$$v^u(a, s) = \left. \frac{\partial U(a, s)}{\partial F} \right|_{F=0} = \frac{t}{E'} \int_0^a K_I(a) \frac{\partial K_I^f(a, s)}{\partial F} da. \qquad (4.2)$$

We now introduce a pair of virtual forces Q at the same locations and the expression for the vertical displacement can be obtained as

$$u^u(a, s) = \left. \frac{\partial U(a, s)}{\partial Q} \right|_{Q=0} = \frac{t}{E'} \int_0^a K_I(a) \frac{\partial K_I^q(a, s)}{\partial Q} da. \qquad (4.3)$$

---

[1] For a 2-D body a point force is defined as a line force normalized by the width in the z-direction.

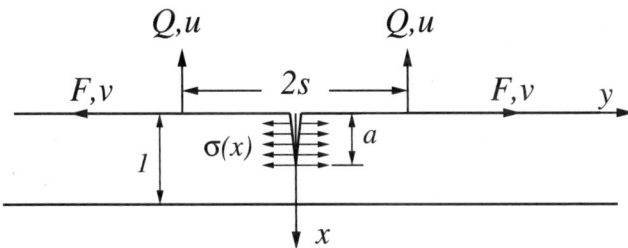

Fig. 4.2. Virtual forces F and Q introduced at locations where displacements are to be determined.

An expression for $K_I(a)$ may be written as

$$K_I(a) = \sqrt{\pi a t}\, f_o(a) \int_0^a f(x,a)\sigma_y(x)dx \qquad (4.4)$$

where $f_o(a)$ and $f(a)$ are given in Appendix A.

Turning to the virtual horizontal force F and vertical force Q, the corresponding normal stresses may be expressed respectively as

$$\sigma_f(s,x) = \frac{F}{\pi t}S(s,x)$$

and

$$\sigma_q(s,x) = \frac{Q}{\pi t}T(s,x)$$

where $S(s,x)$ and $T(s,x)$ are obtained in [36] and are given in Appendix B. The expressions for $K_I^f(a)$ and $K_I^q(x)$ now become

$$K_I^f(a) = F\sqrt{\frac{a}{\pi t}}\, f_o(a) \int_0^a f(x,a)S(s,x)dx \qquad (4.5)$$

and

$$K_I^q(a) = Q\sqrt{\frac{a}{\pi t}}\, f_o(a) \int_0^a f(x,a)T(s,x)dx. \qquad (4.6)$$

Substituting Eqs. (4.4), (4.5) and (4.6) into Eqs. (4.2) and (4.3), we find

$$v^u(a,s) = \frac{t}{E'} \int_0^a \left[a[f_o(a)]^2 \int_0^a f(x,a)\sigma_y(x)dx \int_0^a f(x,a)S(s,x)dx \right] da \qquad (4.7)$$

and

$$u^u(a,s) = \frac{t}{E'} \int_0^a \left[a[f_o(a)]^2 \int_0^a f(x,a)\sigma_y(x)dx \int_0^a f(x,a)T(s,x)dx \right] da$$

(4.8)

Replacing $S(s,x)$ and $T(s,x)$ with $S(s,1-x)$ and $T(s,1-x)$ in Eqs.(4.7) and (4.8), the expressions for displacements $v^l(s,x)$ and $u^l(s,x)$ at the lower edge can be obtained. For a uniform surface stress σ and $\sigma/E' = 1$ displacements $u(s)$ and $v(s)$ are calculated and shown in Figs. 4.3 and 4.4 for different crack sizes.

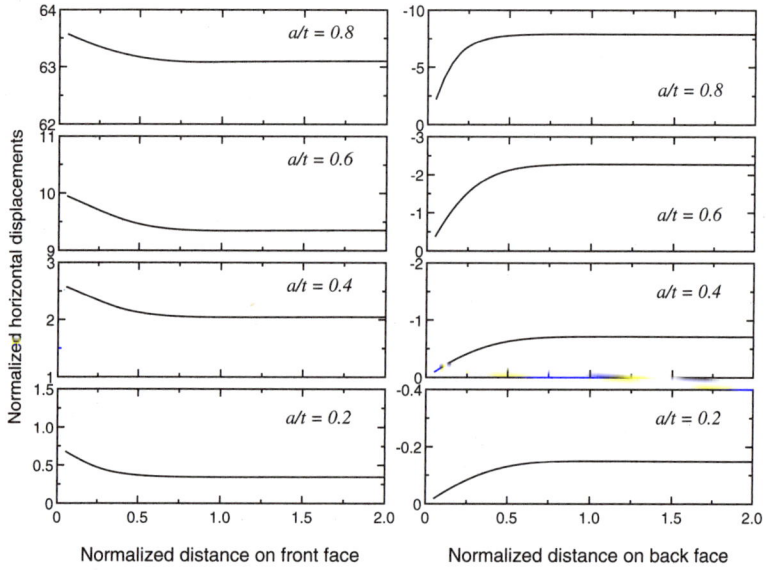

Fig. 4.3. Normalized horizontal displacements on front and back faces.

A postulate made by the authors in [12] states that the displacement at a region more than one thickness away from the crack becomes a rigid body movement. This implies that the displacements are related by

$$(v^u - v^l)/t = u_m/s \qquad for \ s > 1$$

(4.9)

where $u_m = (u^u + u^l)/2$ is the mean vertical displacement at distance s.

To verify this relation, the quantities on the left and right hand sides of Eq. (4.9) are plotted against s in Fig. 4.5 for $a = 0.2, 0.4, 0.6$ and 0.8. It is seen that the two curves indeed merge into one for s less than unity.

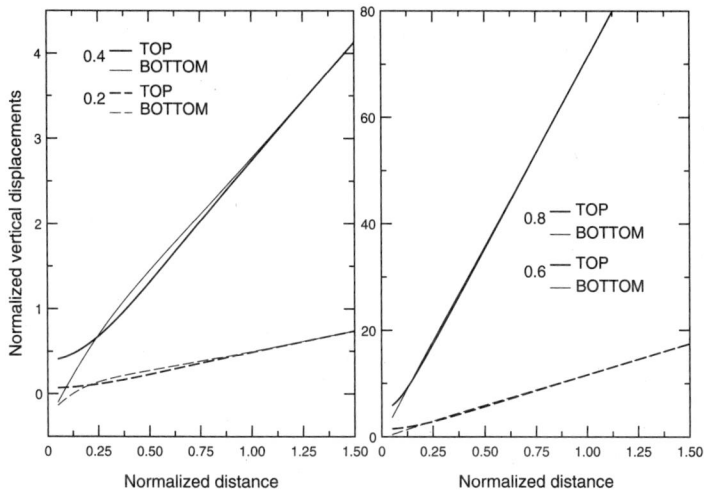

Fig. 4.4. Normalized vertical displacements on front and back faces.

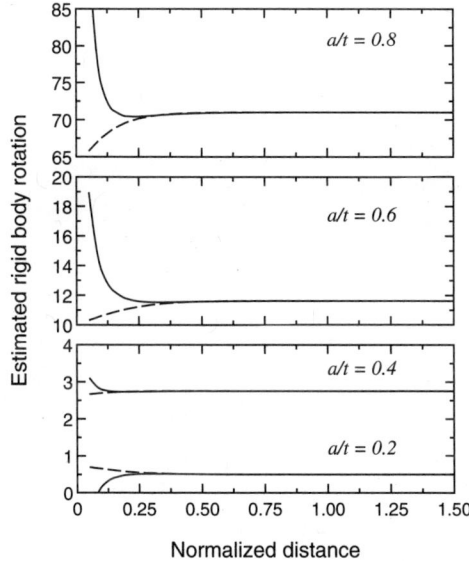

Fig. 4.5. Quantities on both sides of Eq. (4.9) against distance from the crack.

From Eqs. (4.7) and (4.8) we see that the variation of the displacements with distance s is dependent only on the stress field due to the virtual forces F or Q but not on the surface traction on crack faces. Therefore, the deformation due to a non-uniform stress at one thickness away from the crack should also become a rigid body movement.

To determine the normal strain $\epsilon_y^u(y = s)$ on the upper edge produced by the surface traction $\sigma(x)$ on crack faces, we take a derivative of Eq. (4.7) with respect to s. That is,

$$\epsilon_y^u(a, s) = \frac{\partial v^u(a, s)}{\partial s} = \frac{1}{E'} \int_0^a K_I \frac{\partial^2 K_I^f}{\partial F \partial s} da$$

$$= \frac{1}{E'} \int_0^a \left[a[f_o(a)]^2 \int_0^a f(x, a)\sigma(x)dx \int_0^a f(x, a)\frac{\partial S(s, x)}{\partial s}dx \right] da \qquad (4.10)$$

Replacing $S(s, x)$ with $S(s, 1 - x)$ in Eq. (4.10), the expressions for strains $\epsilon_y^l(s, x)$ at the lower edge can also be obtained. Expressions for $S(s, 1 - x)$ and $\partial S(s, 1 - x)/\partial s$ are given in Appendix B.

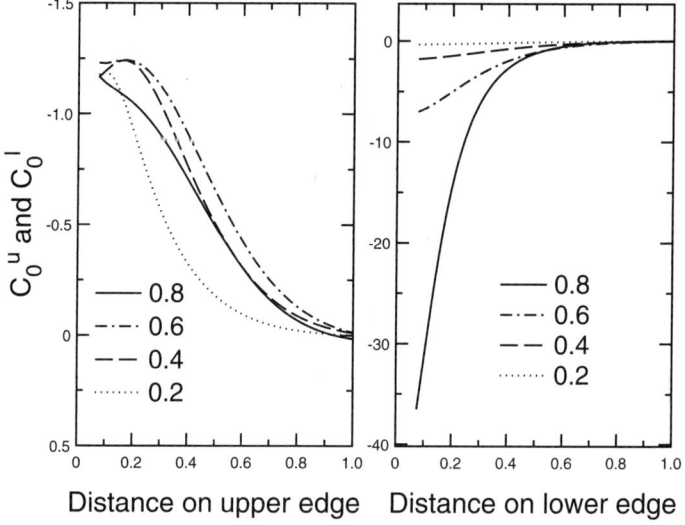

Fig. 4.6. Variation of crack compliances on upper and lower edges of the beam.

The computation of Eq. (4.10) is complicated because the point load solution also requires a semi-infinite integration which may be carried out using Gaussian-Laguerre quadrature. Fortunately, the variation of the stress field due to point forces is smooth and can be approximated accurately by a polynomial series for a given value of s. Since for residual stress measurement we are mostly concerned with the strain at $s = 0$ at the lower edge, an expression using Legendre polynomials in this case may be obtained as

$$\frac{\partial S(s, 1-x)}{\partial s}\bigg|_{s=0} = \frac{12}{\pi} L_1(x) + \sum_{i=2}^{n} b_i \, L_i(x) \qquad (4.11)$$

where b_i is the amplitude factor for the ith order Legendre polynomial and is given in Table B.1. For $n = 15$ Eq. (4.11) leads to an accuracy of better than 10^{-9}. One can readily show that the first term in Eq. (4.11) is exact and the second term results in neither resultant force nor moment. This approach avoids any computational error that leads to a non-zero resultant force and/or moment, which increasingly influences the computed result as crack size increases. A subroutine written in C programming language is given in Appendix C.2 for calculating the summation of Legendre polynomials directly.

The stress distribution $\sigma_y(x)$ in Eq. (4.10) is often expressed as a nth order polynomial series

$$\sigma_y(x) = \sum_{i=0}^{n} \sigma_i P_i(x) \qquad (4.12)$$

where σ_i is an amplitude factor to be determined for the ith order polynomial $P_i(x)$. Substituting Eq. (4.12) into Eq.(4.10), the unit response on the upper edge of the beam to the ith order stress field $P_i(x)$ becomes

$$C_i^u(a, s) = \int_0^a \left[a[f_o(a)]^2 \int_0^a f(x, a) P_i(x) dx \int_0^a f(x, a) \frac{\partial S(s, x)}{\partial s} dx \right] da \qquad (4.13)$$

in which $C_i^u(a, s)$ is referred to as the crack compliance function. Equation (4.10) can now be rewritten as

$$\epsilon_y^u(a, s) = \sum_{i=0}^{n} \frac{\sigma_i}{E'} C_i^u(a, s) \qquad (4.14)$$

For a uniform stress, i.e. $P_0(x) = 1$ the compliance function $C_0^u(a, s)$ and $C_0^l(a, s)$ are calculated and shown in Fig. 4.6 for different crack sizes. It is seen that the strain on the lower face reaches a maximum at the location directly opposite the crack. As will be shown in the next section, this location is ideal for the measurement of normal residual stresses through-the-thickness. The rapid-increasing of the strain gradient on the lower face with the depth of cut suggests that a shorter gage length will lead to a more sensitive measurement of the deformation.

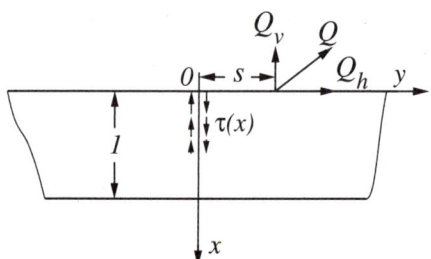

Fig. 4.7. An edge-cracked beam with shear stress on crack faces.

4.2.2 Shear Stresses on Crack Faces

For the same edge-cracked beam shown in Fig. 4.1 we now apply a shear stress $\tau(x)$ to the faces of the crack, as shown in Fig. 4.7. To obtain the displacement at a distance s from the crack, we introduce a virtual force Q at that location. The change of strain energy in the region of $y \leq 0$ due to the presence of the crack is then related to the mode II stress intensity factors K_{II}, K_{II}^h and K_{II}^v produced by $\tau(x)$ and the horizontal and vertical components Q_h and Q_v of the virtual force Q, i.e.

$$U(a, s) = \frac{t}{2E'} \int_0^a \{[K_I^q(a)]^2 + [K_{II}(a, s) + K_{II}^h(a, s) + K_{II}^v(a, s)]^2\} da \quad (4.15)$$

where K_I^q is the mode I stress intensity factor due to virtual force Q.

Using Castigliano's theorem, the expressions for the horizontal and vertical displacements at distance s from the crack can be obtained as

$$v^u(a, s) = \frac{\partial U(a, s)}{\partial Q_h}\bigg|_{Q_h=0} = \frac{t}{E'} \int_0^a K_{II}(a) \frac{\partial K_{II}^h(a, s)}{\partial Q_h} da \quad (4.16)$$

$$u^u(a, s) = \frac{\partial U(a, s)}{\partial Q_v}\bigg|_{Q_v=0} = \frac{t}{E'} \int_0^a K_{II}(a) \frac{\partial K_{II}^v(a, s)}{\partial Q_v} da \quad (4.17)$$

Taking a derivative of Eq. (4.16) with respect to s leads to an expression for the normal strain $\epsilon_y(s, a)$

$$\epsilon_y^u(a, s) = \frac{\partial v^u(a, s)}{\partial s} = \frac{1}{E'} \int_0^a K_{II} \frac{\partial^2 K_{II}^h}{\partial Q_h \partial s} da \quad (4.18)$$

It is seen that Eqs. (4.16), (4.17) and (4.18) are in the same form as those for the normal stress $\sigma_y(x)$ but with corresponding mode I stress intensity factors replaced by the mode II stress intensity factors.

Substituting the solutions of K_{II}, K_{II}^h and K_{II}^v given in Appendix A into Eqs. (4.16), (4.17) and (4.18), the displacements and strain produced by the shear stress on crack faces can be calculated. For a parabolic shear stress distribution $\tau(x) = \tau_o(1 - x)x$ on crack faces and $\tau_o/E' = 1$, the horizontal and vertical displacements are shown in Figs. 4.8 and 4.9 respectively. It is seen that the horizontal displacements on the upper and lower faces vanish rapidly as s approaches unity while the vertical displacements become essentially a uniform movement at $s = 1$. The corresponding distribution of $\epsilon(s, a)$, as shown in Fig. 4.10, also vanishes on both faces for $s > 1$. This implies that the deformation due to shear stress on crack faces reduces to a rigid body movement in the vertical direction when $s > 1$.

The fact that the elastic deformation vanishes at a distance larger than one thickness from the crack plane is consistent with the Saint-Venant's principle. This is because the stresses produced by loading on crack faces always satisfy the force and moment equilibrium.

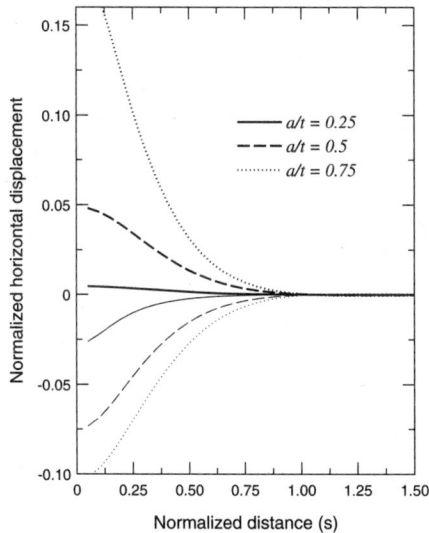

Fig. 4.8. Horizontal displacements on upper (thick line) and lower (thin line) faces due to shear stresses.

It is worth noting that the strain due to a shear stress shown in Fig. 4.10 is always zero at the location directly opposite the crack. A comparison with the strain due to the normal stress shown in Fig. 4.6 indicates that *strain measured at this location corresponds only to the normal stress and will not be influenced by the presence of any shear stress on the plane of the crack.*

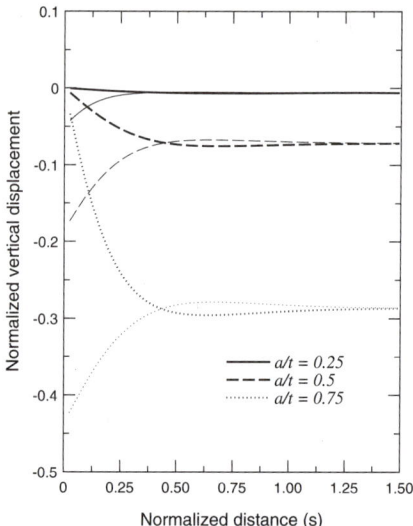

Fig. 4.9. Vertical displacements on upper (thick line) and lower (thin line) faces due to shear stresses.

4.3 An Edge-Cracked Circular Body

The use of Castigliano's theorem to obtain crack compliances applies to any geometries for which the solutions of stress intensity factors are available for all crack sizes. In this section the crack compliance functions for an edge-cracked circular body shown in Fig. 4.11 will be obtained. To simplify the formulation below, we take the diameter D as unity. Thus, for a body with $D \neq 1$ the dimensions a, x and s, shown in Fig. 4.11 can be considered as those normalized by its diameter.

When a surface traction $\sigma(x)$ is applied to the crack faces, the deformation or the change of strain along the perimeter can be obtained using Castigliano's theorem and stress intensity factor solutions. Following the same approach used earlier, the change in hoop strain may be expressed as

$$\epsilon_\theta(a,s) = \frac{1}{E'} \int_0^a K_I \frac{\partial^2 K_I^f}{\partial F \partial s} da \qquad (4.19)$$

in which $E' = E$ for a disk and $E' = E/(1 - \nu^2)$ for a cylinder. The stress intensity factors K_I and K_I^f are respectively for the loading on the crack faces and for a pair of virtual forces F acting symmetrically about the crack at a tangential distance s shown in Fig. 4.11.

An approximate solution of K_I given in Appendix D can be written as

$$K_I = \frac{1.12\sqrt{\pi a D}}{(1-a)^{3/2}} \int_0^a F_{disk}(x,a)\sigma(x) dx \qquad (4.20)$$

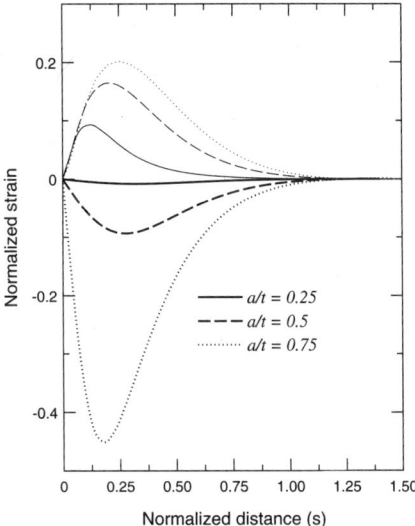

Fig. 4.10. Normal strains on upper (thick line) and lower (thin line) faces due to shear stresses.

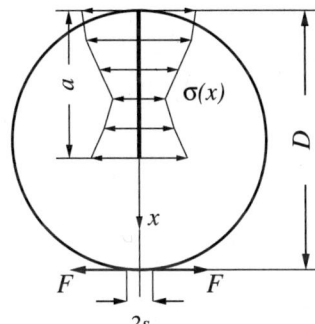

Fig. 4.11. An edge-cracked circular body subjected to a surface traction $\sigma(x)$ and virtual forces F.

where

$$F_{disk}(x,a) = -(\frac{2}{\pi})\frac{\partial}{\partial x}\{(1-x)^2 cos^{-1}[\frac{x(1-a)}{(1-x)a}]H(x,a)\}$$

and

$$H(x,a) = 1 - \frac{3}{28}\frac{x}{a}(1\ -\ a)$$

Taking the limit $s \to 0$, the second term of the integrand in Eq. (4.19) is shown in Appendix E to reduce to a very simple form. That is,

$$\frac{\partial^2 K_I^f}{\partial F \partial s}\Big|_{s=0} = \frac{4.48}{(1-a)^{3/2}D}\sqrt{\frac{a}{\pi D}}$$

which, when substituted into Eq. (4.20), leads to

$$\epsilon(a) = \frac{5.0176}{E'}\int_0^a \frac{ada}{(1-a)^3}[\int_0^a F_{disk}(x,a)\sigma(x)dx] \qquad (4.21)$$

When a stress distribution given in Eq. (4.12) is substituted into Eq. (4.21), the ith order crack compliance functions may be expressed as

$$C_i(a) = 5.0176\int_0^a \frac{ada}{(1-a)^3}[\int_0^a F_{disk}(x,a)P_i(x)dx] \qquad (4.22)$$

Since residual stresses in a body without any external loading always satisfy equilibrium conditions, Legendre polynomials instead of power functions are often used for $P_i(x)$. In most textbooks Legendre polynomials are defined over a domain $-1 \le x \le 1$. For representation of residual stress distributions, however, it is more convenient to express Legendre polynomials over a domain $0 \le x \le 1$, which corresponds to a distance normalized by the thickness. In this case the general form of Legendre polynomials [70] becomes

$$L_i(x) = \frac{(\frac{d}{dx})^n[(x^2-x)^n]}{n!} \qquad for \ \ 0 \le x \le 1 \qquad (4.23)$$

from which the first eight terms are

$$L_0 = 1 \qquad L_1 = 2x - 1$$

$$L_2 = 6x^2 - 6x + 1$$

$$L_3 = 20x^3 - 30x^2 + 12x - 1$$

$$L_4 = 70x^4 - 140x^3 + 90x^2 - 20x + 1$$

$$L_5 = 252x^5 - 630x^4 + 560x^3 - 210x^2 + 30x - 1$$

$$L_6 = 924x^6 - 2772x^5 + 3150x^4 - 1680x^3 + 420x^2 - 42x + 1$$

$$L_7 = 3432x^7 - 12012x^6 + 16632x^5 - 11550x^4 + 4200x^3 - 756x^2 + 56x - 1$$

Readers can verify easily that all terms with $i > 1$ satisfy both the force and moment equilibrium conditions. For this reason the uniform and linear terms in Eq. (4.23) can be omitted for approximation of a residual stress distribution through the thickness of a beam or disk. This leads to a truncated Legendre polynomial series. As will be shown in Chapter 6, the numerical computation for least squares fit becomes more stable when a truncated Legendre polynomial series is used. Figure 4.12 shows the variation of $C_i(a)$ calculated for $\sigma(x) = L_2$ to L_5.

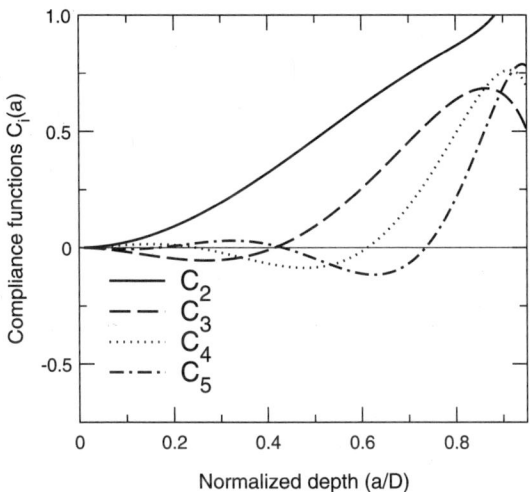

Fig. 4.12. Crack compliance functions for $L_2(x)$ to $L_5(x)$.

4.4 A Thin-Walled Cylinder With a Circumferential Crack

Figure 4.13 shows a thin-walled cylinder of mean radius R and wall thickness t with an arbitrary axisymmetric axial residual stress which is symmetrical about the plane $z = 0$. When a complete circumferential crack is introduced on the plane of symmetry, the deformation could be computed by using a 3-D axisymmetric finite element program. However, equivalent results can be obtained by two much simpler approaches. In the first approach a small annular element containing the crack is first conceptually separated from the cylinder [11]. Then the continuity of the rotation between the element and the rest of the cylinder leads to the crack compliance function for the hoop strain. The second approach, which gives the identical result as the first one, will be presented here.

We start with applying the residual stress on the crack faces. If the cylinder is not constrained in the axial direction, the stress produces only a double moment M on the plane of the crack, which causes a discontinuity in rotation θ at $z = 0$ as shown in Fig. 4.14. The corresponding radial displacement $w(z)$ for a thin-walled cylinder [62] is then given by [11]

$$w(z, \frac{a}{t}) = \frac{2DRM[\sin(\beta z) - \cos(\beta z)]e^{-\beta z}}{Et^2} \tag{4.24}$$

where $D = [3(1 - \nu^2)]^{1/2}$ and $\beta = (D/Rt)^{1/2}$. Taking a derivative of Eq. 4.24 with respect to z gives the rotation

$$\theta(z, \frac{a}{t}) = \frac{4D\beta RMe^{-\beta z}\cos(\beta z)}{Et^2} \tag{4.25}$$

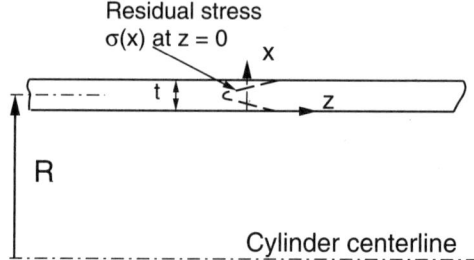

Fig. 4.13. A thin-walled cylinder with an axisymmetric residual stress.

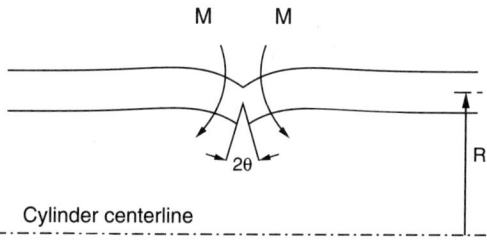

Fig. 4.14. A double moment caused by loading on crack faces.

We now cut out a longitudinal strip along the axial direction from the cylinder shown in Fig. 4.15. The moment M acting on the crack plane is thus released. The change of the rotation can be determined by applying an opposite moment to the strip and the rotation θ_s at $z = 0$ becomes

$$\theta_s = \theta(0, \frac{a}{t}) + M\omega(\frac{a}{t}) \tag{4.26}$$

where the compliance of an edge-cracked beam, $\omega(a/t)$, due to the moment is given by

$$\omega(\frac{a}{t}) = \frac{12(1 - \nu^2)S(\frac{a}{t})}{Et^2}$$

in which[123]

$$S(\frac{a}{t}) = (\frac{a/t}{1 - a/t})^2[5.39 - 19.69\frac{a}{t} + 37.14(\frac{a}{t})^2 - 35.84(\frac{a}{t})^3 + 13.12(\frac{a}{t})^4]$$

Because the strip is subjected to only the surface traction on the crack faces, the rotation is also given by

$$\theta_s = \theta^p\left(\frac{a}{t}\right) \tag{4.27}$$

which is the plane strain solution for an edge-cracked beam. Replacing the virtual force F in Eq. 4.2 by a virtual moment M', the rotation is found to be

$$\theta^p(a,s) = \frac{\partial U(a,s)}{\partial M'}\bigg|_{M'=0} = \frac{t}{E'}\int_0^{a/t} K_I(a/t)\frac{\partial K_I^m(a/t,s)}{\partial M'}d(a/t) \tag{4.28}$$

in which

$$\frac{\partial K_I^m(a/t,s)}{\partial M'} = \frac{6\sqrt{\pi a}}{t^2}f_0\left(\frac{a}{t}\right)\int_0^{a/t} f\left(x,\frac{a}{t}\right)(2x-1)dx$$

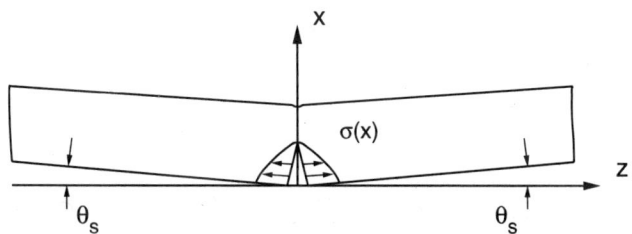

Fig. 4.15. Rotation of an edge-cracked beam.

Combining Eq. (4.26) with Eq. (4.27), the moment can be found as

$$M(a/t) = \frac{\theta^p(a/t)}{\frac{4R\beta D}{Et^2} + \omega(a/t)} = \frac{\theta^p(a/t)E't^2}{12[\frac{1}{\beta t} + S(a/t)]} \tag{4.29}$$

which, when substituted into Eq. (4.24), the radial deflection w at a distance z from the crack plane can be obtained

$$w(z,a/t) = \frac{\theta^p(a/t)R[\sin(\beta z) - \cos(\beta z)]e^{-\beta z}}{2D[\frac{1}{\beta t} + S(a/t)]}. \tag{4.30}$$

In practice the hoop strain is easier to measure than the radial deflection, and a straightforward relationship exists for thin-walled cylinders. That is, for hoop strain measured at the outside surface,

$$\epsilon(z,a/t) = \frac{w(z,a/t)}{R+t/2} = \frac{\theta^p(a/t)[\sin(\beta z) - \cos(\beta z)]e^{-\beta z}}{D[\frac{1}{\beta t} + S(a/t)](2+t/R)} \tag{4.31}$$

For a residual stress distribution expressed by Eq. (4.12), Eq. (4.28) becomes

$$\theta^p\left(\frac{a}{t}\right) = \frac{1}{E'}\sum_{i=0}^{n}\sigma_i\theta_i^p\left(\frac{a}{t}\right)$$

where σ_i is the amplitude factor of the axial stress for the ith term in the series. Equation (4.31) now may be rewritten as

$$\epsilon(z, a/t) = \frac{e^{-\beta z}}{E'}[\sin(\beta z) - \cos(\beta z)]\sum_{i=0}^{n}\sigma_i C_i^a(a/t, R/t) \tag{4.32}$$

where the compliance function C_i^a for the hoop strain is given by

$$C_i^a(a/t, R/t) = \frac{\theta_i^p(a/t)}{D[\frac{1}{\beta t} + S(a/t)](2 + t/R)} \tag{4.33}$$

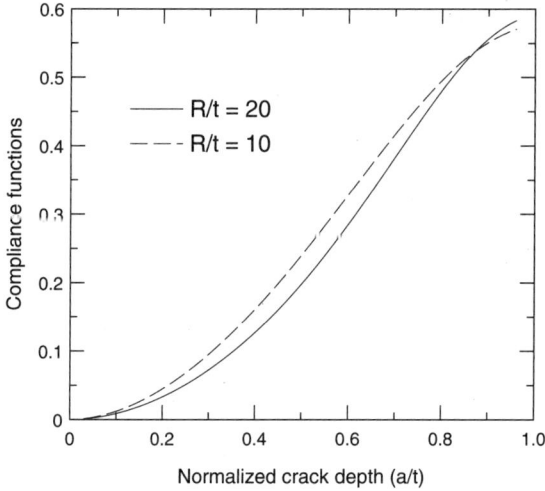

Fig. 4.16. Crack compliance function for a bending stress ($\nu = 0.3$).

Compliance functions given in Eq. (4.33) are based on the solutions for a thin-walled cylinder for which the ratio of the mean radius to wall thickness R/t is equal to or larger than ten. However, the present solution was also found to be in close agreement with the numerical results calculated by finite element method for $R/t = 5.1$ [12]. Figure 4.16 shows the compliance functions for $R/t = 10$ and 20 as a function of a/t for an axisymmetric bending stress on the crack faces.

4.5 A Ring With a Radial Crack

To obtain the compliance functions for a cylinder of mean radius R and wall thickness t, we may introduce a radial crack from either inside or outside wall. Here, we focus our attention on a cut made from the outside the wall. This leads to a configuration shown in Fig. 4.17. Again, instead of resorting to a numerical computation we separate conceptually a small element containing the crack from the cylinder as shown in Fig. 4.17. The stress distribution on the crack faces corresponds to the original residual hoop stress in the uncracked body with an opposite sign. The moment M and force F per unit axial distance are those due to introduction of the crack. Since the cracked element subtends an infinitesimal angle, the hoop strain measured on the outer surface of the cylinder at an angular location ϕ with respect to the crack plane can be expressed as

$$\epsilon(a, \phi) = \frac{6H_0(\bar{R})}{t^2 E'}[M - FR(1 - H(\bar{R})\cos\phi)] \tag{4.34}$$

where $\bar{R} = R/t$ and

$$H_0(\bar{R}) = \frac{1}{3}\frac{\bar{R} - (\bar{R} - 0.5)^2\ln(\frac{\bar{R}+0.5}{\bar{R}-0.5})}{\bar{R}^2 - (\bar{R}^2 - 0.25)^2[\ln(\frac{\bar{R}+0.5}{\bar{R}-0.5})]^2}$$

$$H(\bar{R}) = \frac{(\bar{R} + 0.5)\{[\ln(\frac{\bar{R}-0.5}{\bar{R}+0.5})]^2 - \frac{\bar{R}^2}{(\bar{R}^2-0.25)^2}\}}{[\bar{R} + (\bar{R}^2 + 0.25)\ln(\frac{\bar{R}-0.5}{\bar{R}+0.5})][\frac{\bar{R}}{(\bar{R}-0.5)^2} + \ln(\frac{\bar{R}-0.5}{\bar{R}+0.5})]} \tag{4.35}$$

We now need to derive expressions which relate the hoop stress to the moment M and force F as a function of the crack length. Denoting the rotation and extension produced by the moment and force acting on the ends of the cracked element with superscripts m and f respectively, the rotation θ and extension v at each end of the element can be expressed as

$$\theta = \theta^p - \theta^m + \theta^f \quad , \quad v = v^p - v^m + v^f \tag{4.36}$$

where θ^p and v^p are the rotation and extension due to loading on the crack faces and have been derived using Castigliano's theorem in previous sections. The other four variables in Eq. (4.36) may be expressed as

$$\theta^m = M\lambda^m \qquad \theta^f = F\lambda^f \qquad v^f = F\tilde{\lambda}^f \qquad v^m = M\tilde{\lambda}^m \tag{4.37}$$

The λ's are compliances with extension compliance denoted by a tilde.

By continuity the rotation θ and extension v of the one half of the split cylinder must equal those at the end of the cracked element. Using a subscript s to denote θ and v for one half of the split member in Fig. 4.17, we have

$$\theta = \theta_s^m - \theta_s^f \qquad V = -V_s^f + V_s^m \tag{4.38}$$

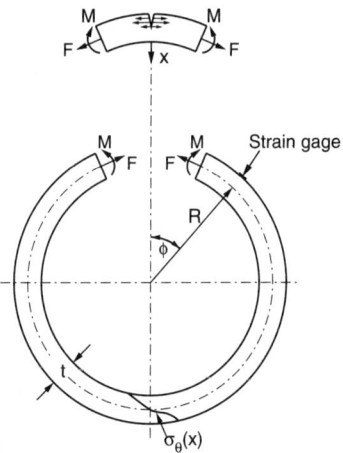

Fig. 4.17. An element containing the crack separated from the ring.

The quantities θ_s^m, θ_s^f, V_s^m and V_s^f are found in texts on elasticity [124]

$$\theta_s^m = M\lambda_s^m = \frac{M}{H_1(\bar{R})}\frac{12\pi R}{(E't^3)}$$

$$\theta_s^f = F\lambda_s^f = \frac{F}{H_1(\bar{R})}\frac{12\pi R^2}{E't^3}$$

$$V_s^f = F\,\lambda_s^f = \frac{F18\pi R^3}{E't^3}\frac{H_1(\bar{R}) + 2H_2(\bar{R})}{3H_1(\bar{R})H_2(\bar{R})} \qquad (4.39)$$

$$V_s^m = M\lambda_s^m = \frac{M}{H_1(\bar{R})}\frac{12\pi R^2}{E't^3}$$

where $H_1(\bar{R})$, $H_2(\bar{R})$ are correction functions for thick-walled cylinders given by

$$H_1(\bar{R}) = 3\{\bar{R}^2 - [(\bar{R}^2 - 0.25)\ln(\frac{\bar{R} + 0.5}{\bar{R} - 0.5})]^2\}$$

$$(4.40)$$

$$H_2(\bar{R}) = 3\bar{R}^3[\ln(\frac{\bar{R} + 0.5}{\bar{R} - 0.5}) - \frac{\bar{R}}{\bar{R}^2 + 0.25}].$$

For R/t ratio larger than 4.5 the difference between Eq. (4.40) and results found in texts on strength of materials [62] becomes less than 0.5%. Substituting Eq. (4.40) into Eq. (4.39), equating Eq. (4.36) and Eq. (4.38) and making use of Eq. (4.12), the moment and force can be found as

$$M = \sum_{i=0}^{n} \sigma_i M_i(\frac{a}{t}); \quad F = \sum_{i=0}^{n} \sigma_i F_i(\frac{a}{t}) \qquad (4.41)$$

in which

$$M_i(\frac{a}{t}) = [F_i^\theta(\frac{a}{t})(\tilde{\lambda}^f + \tilde{\lambda}_s^f) - F_i^v(\frac{a}{t})(\lambda^f + \lambda_s^f)]/(\Delta E')$$

$$F_i(\frac{a}{t}) = [-F_i^v(\frac{a}{t})(\lambda^m + \lambda_s^m) + F_i^\theta(\frac{a}{t})(\tilde{\lambda}^m + \tilde{\lambda}_s^m)]/(\Delta E') \qquad (4.42)$$

with

$$\Delta = (\lambda^m + \lambda_s^m)(\tilde{\lambda}^f + \tilde{\lambda}_s^f) - (\tilde{\lambda}^m + \tilde{\lambda}_s^m)(\lambda^f + \lambda_s^f)$$

The hoop strain given in Eq. (4.34) can now be rewritten as

$$\epsilon(\frac{a}{t}) = \sum_{i=0}^{n} \frac{\sigma_i}{E'} C_i^h(\frac{a}{t}) \qquad (4.43)$$

where

$$C_i^h(\frac{a}{t}) = \frac{6H_0(\bar{R})}{t^2}\{M_i(\frac{a}{t}) - F_i(\frac{a}{t})R[1 - H(\bar{R})\cos\alpha]\}$$

The solution in Eq. (4.43) applies to a cylinder (plane strain) or a ring (plane stress) in which the hoop stress σ_θ is symmetric about the plane of the crack. This implies that the hoop stress does not vary in the axial direction for a cylinder while for a ring the axial stress is zero. Figure 4.18 shows the crack compliance functions at three locations on the outer surface as a function of the normalized crack size when a bending stress is released by cutting.

For a thin-walled non-circular part, the crack compliance functions can be also obtained using the same approach presented here, as demonstrated in [18].

4.6 Discussion

The crack compliance functions presented in this chapter are computed using the stress intensity factor solutions and point force solutions given in Appendices A and B. The integrations in Eq. (4.13), Eq. (4.22), Eq. (4.33), and Eq. (4.43) can be carried out numerically on a personal computer. For simple geometries, such as plates, beams, cylinders, and disks, they are very convenient to implement and have several major advantages:

1. Strain can be calculated at any given depth of cut on any given location on the surface;
2. An arbitrary stress field can be readily specified on the crack faces;
3. The solutions may be used to estimate stress intensity factors directly from measured strain;

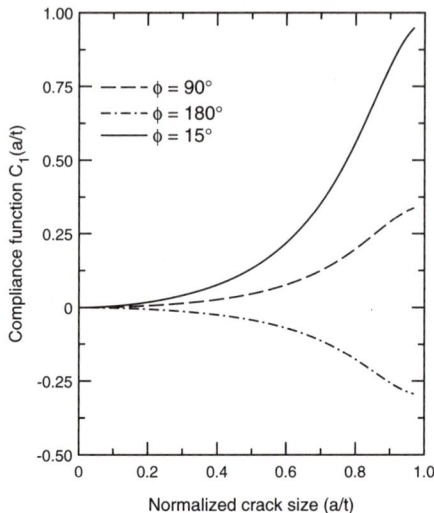

Fig. 4.18. Crack compliance function for a ring with a radial crack subjected to a bending stress.

4. The time required for numerical computation is a small fraction of that required by the finite element method.

Nevertheless, for parts that cannot be approximated by a geometry discussed in this chapter, a more general approach, such as the finite element method (FEM), will be required.

5

Compliance Functions for Through-Thickness Measurement: The FEM Approach

5.1 Introduction

Many configurations which require residual stress measurement differ from those we discussed in the previous chapter and do not have analytical solutions available from fracture mechanics. In this case we can use the finite element method (FEM) [133] to obtain numerically the crack compliance functions for a specimen of a complex geometry. Since this approach allows the compliance functions to be obtained for a crack as well as a slit of finite width, in what follows, we will use the term "compliance functions" interchangeably for a crack and a slot.

5.2 General Consideration in Finite Element Mesh

In a fracture analysis it is the local stresses in the crack tip region that are of interest. For this reason crack tip elements and a fine mesh are usually used in a finite element (FE) computation to model the rapidly varying stress field near the crack tip. For computation of the compliance functions, however, it is the change of displacements at a location away from the crack tip that needs to be obtained. In this case the deformation away from the crack tip, as shown below, is much less influenced by the choice of the mesh near the crack tip.

A typical variation of the stress field near the crack tip shown in Fig. 5.1 may be decomposed into three parts: a mean stress (σ_{y1}), and a bending stress (σ_{y2}) and a non-linear stress distribution (σ_{y3}) which satisfies both force and moment equilibrium. The purpose of using crack elements is to approximate the rapidly varying stress gradient in the third part of the stress near the crack tip. From Saint-Venant's principle, the influence of the third part of the stress (σ_{y3}) vanishes rapidly with the distance from the crack tip. In other words, the displacement at a location remote from the crack tip depends mainly on the first two parts of the stresses shown in Fig. 5.1. Since the mean stress and the resultant bending stress can be computed satisfactorily

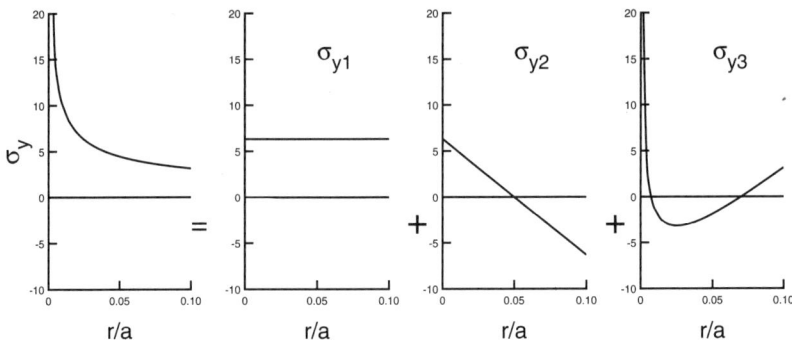

Fig. 5.1. Stress near the crack tip decomposed into: (a) a uniform stress, (b) a bending stress and (c) a nonlinear stress containing the singularity.

with regular elements, crack elements at the crack tip are no longer needed for computation of the displacements at locations remote from the crack tip. This observation is consistent with the fact that the deformation away from the crack tip computed with regular elements has been used to obtain accurate results for J-integral [108] and stress intensity factors [27] in fracture analysis.

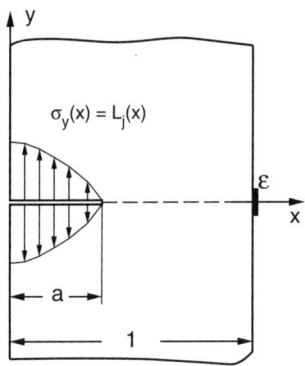

Fig. 5.2. An edge-cracked plate used in the convergence study.

Like any other numerical methods, the accuracy of the FEM approach should be assessed by carrying out a convergence study. A sample geometry of an edge-cracked plate shown in Figure 5.2 is used in this study. For symmetry only one half of the body needs to be modeled, and the crack surface is represented by the unconstrained nodes on the crack plane. As we have shown earlier, the deformation due to a crack is mostly confined within a region a

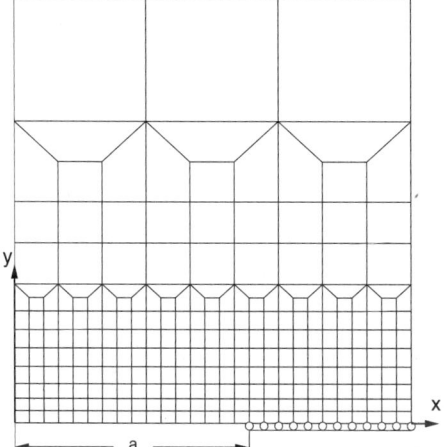

Fig. 5.3. A finite element mesh used to model an edge-cracked plate.

distance about 50% of the thickness from the plane of the crack. At a distance about one thickness the deformation reduces essentially to a rigid body translation and rotation. Therefore, we can use increasingly larger elements in regions away from the crack. Since the change of strain on the back face opposite the crack is of interest, a finer mesh should be used along the entire uncracked ligament. To simplify the modelling of a crack of progressively increasing depth, we also want the elements on the crack plane to be evenly spaced. Based on these considerations, we may construct a mesh as shown in Fig. 5.3, which uses a reduction ratio of three to increase the size of elements away from the crack. Eight-node linear-strain elements are chosen for the study because they lead to a more accurate result than 4-node constant-strain elements. Next, we carry out a series of computations to determine how many elements on the crack plane are required to achieve a sufficient accuracy for the deformation on the back face. The number of elements on the plane of crack is chosen to be respectively 54, 108, 216 and 432, which are all multipliers of three and lead to a very rapid reduction in number of elements away from the crack.

The first sample stress distribution is a uniform stress, i.e., $\sigma = 1$, for which K_I solution given by Eq. 4.4 [111] is accurate within 1%. Normal strains on the back face directly opposite the crack is calculated for $\sigma/E' = 1$. Computation stops when the number of elements in the uncracked ligament reduces to 16, 8, 4 and 2 for $m = 432$, $m = 216$, $m = 108$ and $m = 54$ respectively. Figure 5.4 shows a comparison of the results by finite element computations and by the LEFM solution, Eq. (4.10). For a deep crack with $a/t > 0.8$ the agreement between the two approaches is seen to improve as the number of elements increases and for $m = 432$ the difference at $a/t = 0.96$ decreases to less than 0.7%. For $m = 54$ the difference is about 3% for $a/t = 93\%$. For higher order

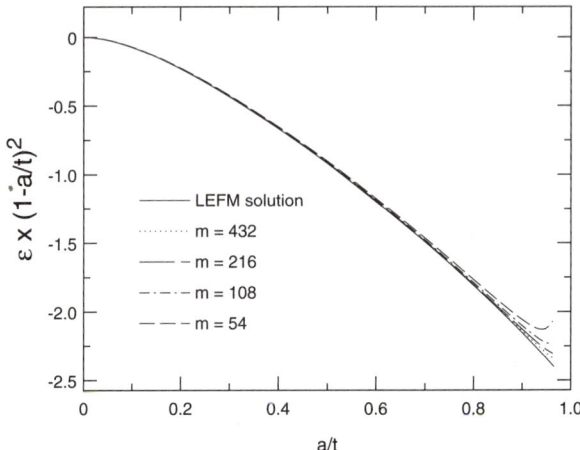

Fig. 5.4. Compliance functions obtained by Eq. (4.10) and FEM for a uniform stress on an edge-cracked plate.

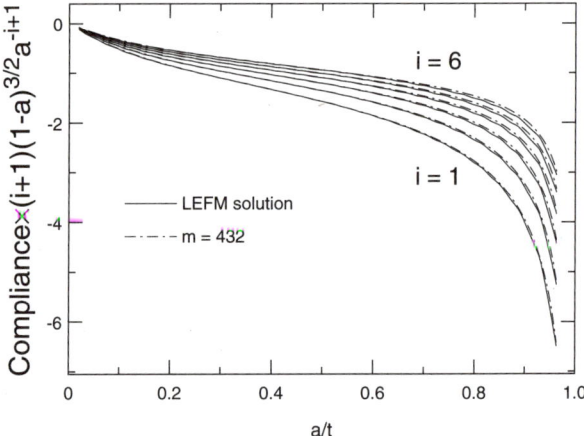

Fig. 5.5. Same as Fig. 5.4 for stresses, $\sigma(x) = x^i$ with $i = 1$ to 6.

stresses the comparison is given in Fig. 5.5 for $m = 432$. The results show that good agreement can be achieved equally for higher order functions.

Since for a deep crack the deformation in the remaining ligament due to a residual stress distribution is much less pronounced than that produced by stresses that result in a net force and/or a net moment, it is more realistic to consider a stress distribution that satisfies both force and moment equilibrium. Thus, the next sample stress field is a second order Legendre polynomial given by

$$L_2(x) = 6x^2 - 6x + 1.$$

for which Eq. (4.10) only leads to an approximate solution.

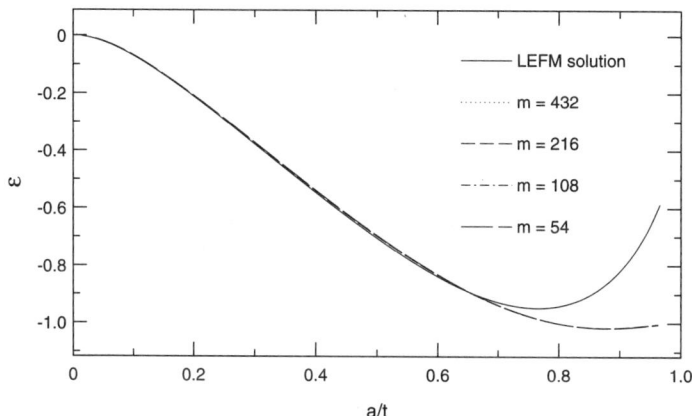

Fig. 5.6. Same as Fig. 5.4 for a 2nd order Legendre polynomial.

Figure 5.6 shows the results by finite element computations and by LEFM or Eq. (4.10). For all crack sizes the agreement among the FE results is excellent. However, they differ considerably from Eq. (4.10) for $a/t > 0.7$. To

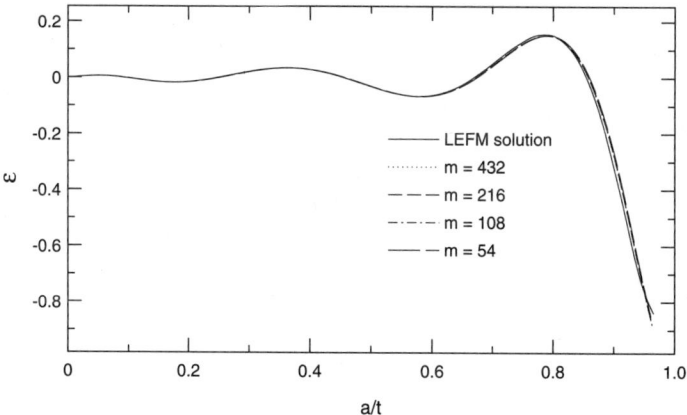

Fig. 5.7. Same as Fig. 5.4 for a 7th order Legendre polynomial.

investigate further the deformation computed by FEM and Eq. (4.10), residual stresses presented by Legendre polynomials of orders higher than 2 are also chosen as the sample stress fields. It is found that the difference in results by FEM and LEFM decreases as the order increases. When the order equals 7, the difference at $a/t = 0.96$ reduces to less than 4%. Figure 5.7 shows the results for the 7th-order Legendre polynomial. For all crack sizes the results by the FEM are seen to agree closely with that by Eq. (4.10).

The most useful finding from the convergency study, Fig. 5.6 and 5.7, is that the accuracy of the computed deformation on the back face due to releasing residual stresses (Legendre polynomials of order 2 or higher) is little influenced by the choice of the mesh when the number of elements on the crack plane is 54 or larger. In practice, the choice of the mesh usually depends upon the largest depth of the residual stress measurement. For $a/t \leq 0.96$ the optimum number of elements appears to $m = n \times 27$ with $n > 1$. This choice will lead to a rapid reduction of the element number at the region away from the crack tip and yet gives a consistent result. The limited deformation found in the remaining ligament for a deep crack subjected to residual stresses also suggests that the corresponding compliances can be obtained with a coarser element mesh than those required for other types of loading that will lead to an unbounded deformation in the remaining ligament.

The most surprising finding of all, of course, is the considerable difference between the results by FEM and the LEFM solution for a deep crack subjected to a residual stress represented by a second order Legendre polynomial. A probable explanation may be as follows. A very good LEFM solution may still have an error about $0.5 - 1\%$ for a regular loading, which is considered sufficient for most fracture problems. A small error of this magnitude becomes enormous when compared with the limited deformation due to residual stress. For example, for $a/t = 0.8$, the deformation due to a uniform stress is about 46 times that due to a residual stress expressed by $L_2(x)$, and an error of 0.5% in the first case is equivalent to an error of 23% in the second case. Thus, it is expected that a very small error in a LEFM solution would be sufficient to cause some variations in compliances computed for a deep crack subjected to a self-balanced load. On the other hand, the finite element computation appears to give more consistent results for a deep crack subjected to a self-balanced load because the deformation of the remaining ligament is sufficiently bounded. A definite answer to which approach is more accurate will have to wait until an exact solution becomes available from fracture mechanics. An immediate question arises: Given the uncertainty in the computed compliance functions for deep cracks, which method is more preferable? It appears that the finite element method is more preferable given its consistency in computing compliances for residual stresses. Additional discussion of this question will be delayed until the next chapter. There we will demonstrate that the difference between the two approaches, fortunately, can be significantly reduced, for example, when a Chebyshev polynomial series is used in computation of the compliances by the LEFM solutions.

5.3 Typical Geometries Analyzed by the FEM

5.3.1 An Edge-Crack at a T-Joint Weld or a Fillet Weld

Welding is one of the most commonly used processes in construction and manufacture [77]. The geometry of a welded part is usually more complicated than those that were covered earlier. Useful examples include a T-joint weld between two plates shown in Fig. 5.8-a or a fillet weld shown in Fig. 5.8-b. The shape of the joint varies with the dimension of the plates and material properties of the two plates may be different. We want to measure the residual normal stress distribution on a plane perpendicular to the surface of the weld region. In this case an analytical solution relating the deformation to

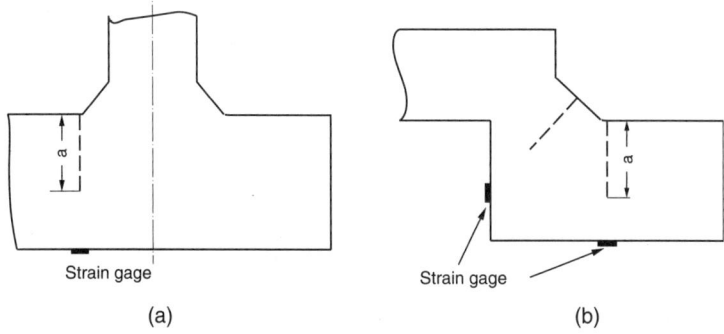

Fig. 5.8. (a) A T-joint weld or (b) a fillet weld between two plates.

stresses on the crack faces is not available. To solve this problem, a hybrid approach combining the solutions for an edge-cracked beam with finite element computations for the welded attachment had been used to obtain the crack compliance functions for a T-joint [25]. A reason for taking this approach was to reduce the amount of computation required by the finite element method. Thanks to the tremendous increase of the computing power in the past two decades, most linear elastic 2-D FE computations can now be carried out on a personal computer. Thus, the reason for using the hybrid approach is much less appealing and a complete finite element analysis makes the formulation of the problem much more general.

We first consider a cut introduced on a plane near the toe of the weld to measure the normal residual stress distribution through the thickness of the plate. The asymmetric geometry of the weld implies that both normal and shear residual stresses may exist on the plane. To minimize the influence of

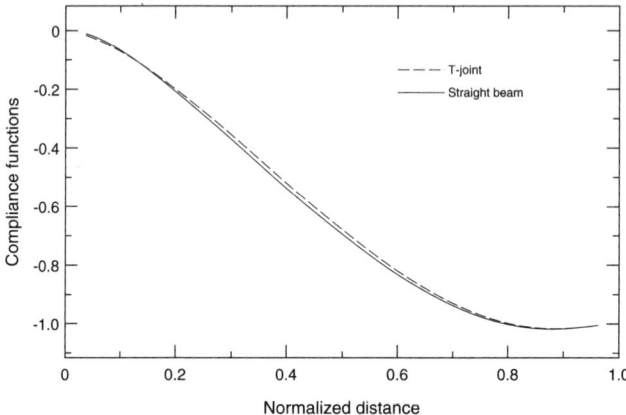

Fig. 5.9. Compliance functions for a cut on a plane near the toe of the T-joint weld.

the shear stress on the measurement of the normal stress, the location on the back face directly opposite the crack is chosen for strain measurement.

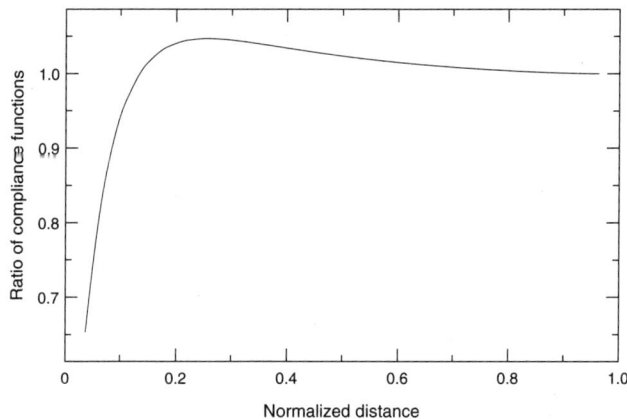

Fig. 5.10. Ratio of the Compliance functions for a cut on a plane near the toe of the T-joint weld and a cut on a straight beam.

The compliance functions are computed for Legendre polynomials and are plotted in Fig. 5.9. As a comparison, the corresponding compliance functions for a plate of uniform thickness are also plotted. The ratio of compliance functions for the two configurations is plotted in Fig. 5.10. It is seen that the difference reaches 70% for a very short crack and decreases rapidly to about 5% or less for $a/t > 0.2$. This result is in close agreement with that obtained by the alternative hybrid approach [25].

To illustrate the flexibility of FEM, we now introduce a cut on a plane normal to the surface of a fillet weld as shown in Fig. 5.8-b. This configuration is obviously too complicated to be handled even by the hybrid approach. Recall

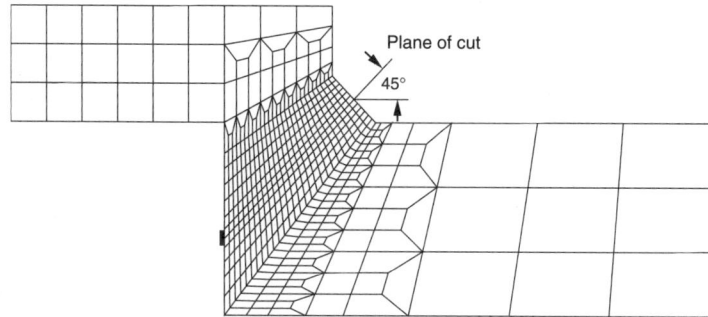

Fig. 5.11. FE mesh used to obtain compliance functions for a 45° cut on the fillet weld.

that in Section 4.2.2 the optimum location for measurement of normal stress in the presence of shear stresses was found to be on the back face opposite the crack for a straight beam. However, the cut now is in an oblique angle of 45° with the back face, and we need to re-examine the influence of the shear stress released along the faces of the cut on the strains measured on that location.

The fillet weld illustrated in Fig. 5.8-b is modelled by a finite element mesh shown in Fig. 5.11. For comparison the average strains over two elements containing the crack plane are computed for a uniform normal stress and a shear stress distribution which leads to the same resultant force as the normal stress. Since the shear stress vanishes at the free surfaces, it is represented by a parabolic distribution for which the maximum magnitude is 50% larger than that of the normal stress. The results are shown in Fig. 5.12. The strain computed on the back face due to the shear stress distribution is seen to be less than 24% of that due to the normal stress. If the maximum magnitude of the residual shear stress is no more than one third of that of the residual normal stress on a plane with an angle less than 45° to the back face, the maximum error in the residual normal stress estimated in the presence of the shear stress is expected to be well within 5% and can be reduced further by using a least squares fit.

5.3.2 A Slot of Finite Width in a Thin Specimen

For through-the-thickness measurement, we have approximated a cut of finite width by a crack. For a thin specimen, however, the width of the cut is no

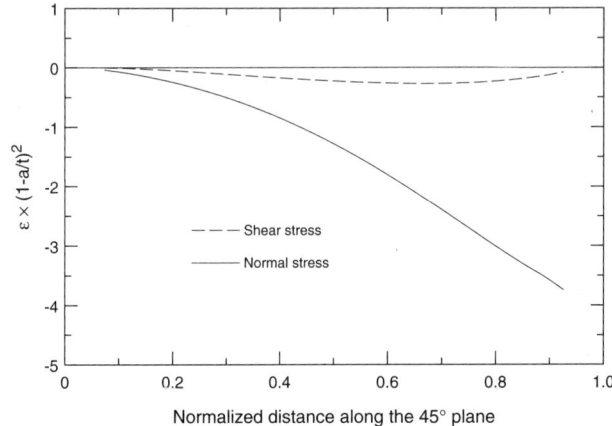

Fig. 5.12. Normalized strains computed for a normal or shear stress acting on the 45° cut.

longer very small relative to the thickness, and the geometry of the slot has to be taken into account in computing the compliance functions.

Depending upon the means of cutting processes, the bottom of the cut may have a variety of geometries. Here we are mainly concerned about a cut with a semi-circular bottom since a cut of progressively increasing depth is most conveniently introduced in measurements by wire electrical discharge machining (wire EDM). Nevertheless, the procedure discussed and results obtained below should be applicable to a cut of other geometries.

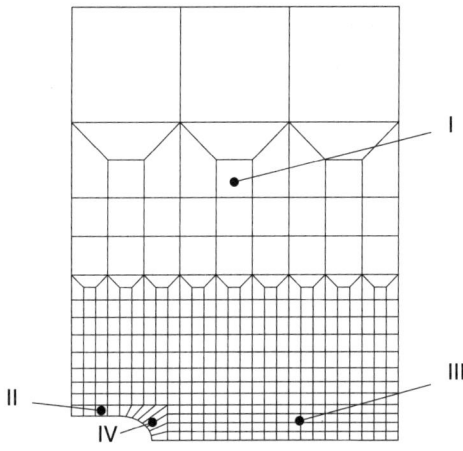

Fig. 5.13. A body with a cut of semi-circular bottom modelled by a FE mesh consisting of four parts.

A typical mesh used for a cut of semi-circular bottom is shown in Fig. 5.13. Unlike a crack for which an increment is achieved by releasing nodes ahead of the crack tip, an advance of the cut has to be modelled by moving the mesh at the bottom of the cut one step at a time. To simplify the process of re-meshing, the whole element mesh may be divided into four parts as denoted in Fig. 5.13 by I though IV. Part I contains the body one element layer above the cut. It is invariant with the advance of the cut and needs to be generated separately only once for cuts of different depth and geometries. Part II is a layer of elements covers the straight side of the cut. The number of elements in this part increases linearly with the advance of the cut. For each cut increment the coordinates of new nodes can easily be generated by a linear interpolation. The third part of the mesh contains the most area of the remaining ligament with a height proportional to the width of the cut. The number of element layers in this part is a linear function of the ratio of the element size to the half width of the cut. As the depth of cut increases, the number of elements decreases linearly. The last part consists of the elements along the circular bottom. Since the geometry and the number of elements remains unchanged for all the increments of the cut, the new coordinates of the element nodes can readily be generated by a linear translation from previous locations.

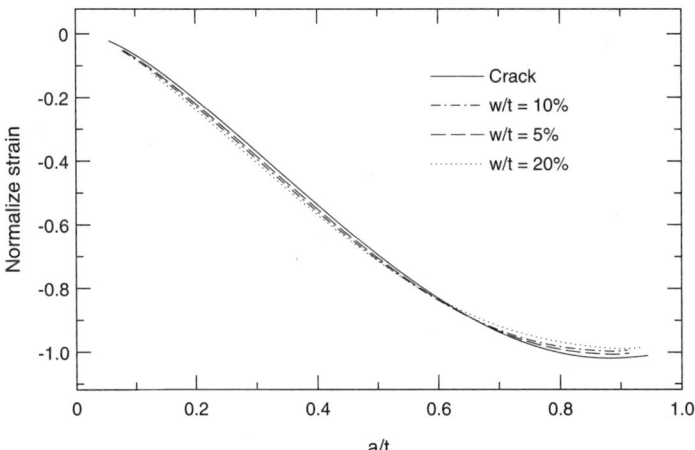

Fig. 5.14. Comparison of the compliance functions for a crack and cuts of finite width.

We are now ready to consider a numerical example for a part with a slot of finite width. Taking the width of the slot as 5%, 10% and 20% of the thickness respectively, the compliances of the slot for a stress given by a second order Legendre polynomial are computed by FEM. A comparison with the compliance for a crack is shown in Fig. 5.14 for $0.08 < a/t < 0.91$. The compliances for the crack and the slot for the three widths are seen to be very similar, and the difference between the two decreases with the width of the cut. Thus, when the width of the slot is less than 3% of the thickness, crack compliances are expected to give a satisfactory approximation.

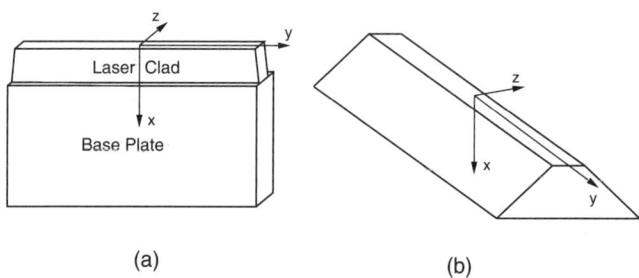

(a)   (b)

Fig. 5.15. (a) A laser-cladded plate and (b) a tapered beam with residual stresses produced by heat treatment.

5.3.3 The Use of 2-D FEM for 3-D Problems

Our adventures with compliance functions have so far been limited to 2-D geometries with a uniform width in the z-direction. In this section we will demonstrate that with a little additional effort compliance functions can also be obtained for parts with a varying width. Figure 5.15 illustrates two configurations for which residual stress measurements need to be carried out. The first configuration arises when multiple layers of laser cladding are deposited on the top surface of a standing plate. During cooling the width of the clad shrinks and becomes less than that of the plate. The second configuration represents a tapered beam with residual stresses being produced by heat treatment or bending. In both cases the variations of the longitudinal residual stresses in the x-direction are of interest and the stress variations in the z-direction are assumed to be uniform.

The geometry of the laser clad plate can be modelled by a 3-D element mesh using different widths. However, a 3-D mesh is more complicated to generate and the computation time is typically more than a hundred times that required for a 2-D FE computation. Therefore, it is very desirable to be able to use a 2-D FE computation to obtain the crack compliance functions for the 3-D bodies shown in Fig. 5.15.

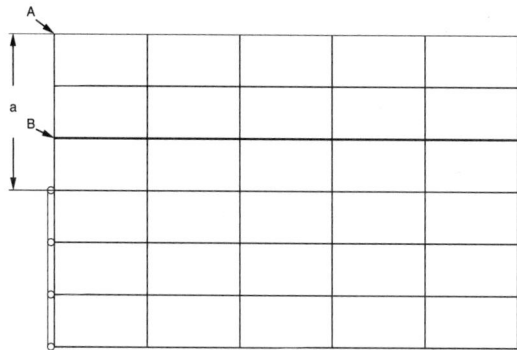

Fig. 5.16. A 2-D FE mesh using different elastic moduli to model width variation of the laser-clad plate.

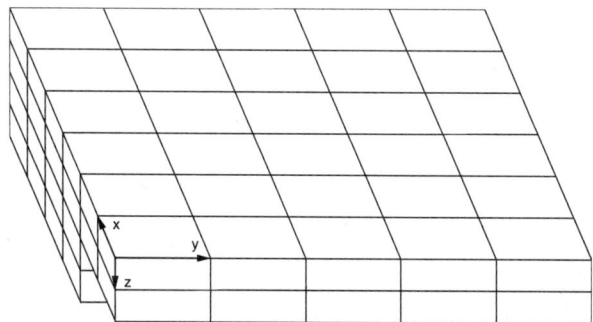

Fig. 5.17. A 3-D finite element mesh with the original material properties of laser-clad plate.

First, we recall that the compliance of a body with a uniform width W and elastic modulus E varies linearly with $1/W$ and $1/E$. Thus, the compliance remains the same if W is scaled down the same amount as E is scaled up. Based on this idea, a simple 2-D element mesh is generated for one half of the laser clad and the plate as shown in Fig. 5.16. The effect of different widths W_c for the clad and W_p for the plate is modelled by using different elastic moduli E_c and E_p according to the ratio of the widths of the clad and the plate. That is,

$$\frac{E_c}{E_p} = \frac{W_c}{W_p}$$

Also the stress σ on the faces of the cut is specified such that the resultant force is the same as that in the 3-D model. This leads to

$$\frac{\sigma_c}{\sigma_p} = \frac{W_p}{W_c}$$

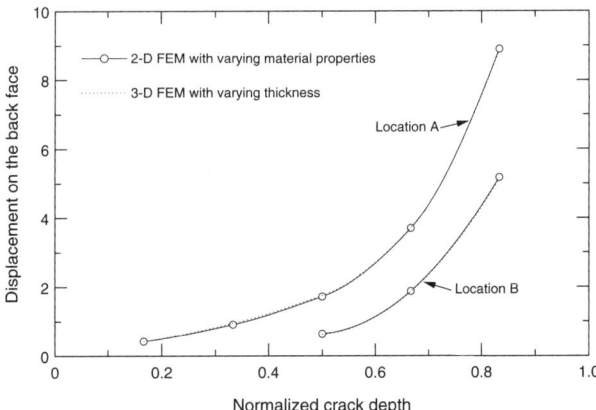

Fig. 5.18. Comparison of the results obtained using either a 3-D or 2-D computation. .

A crack is simulated by releasing nodes along the plane of symmetry. The strains on the back face opposite the crack are computed for a point force applied at the mouth of the crack (location A) or at the interface (location B). For comparison the strains at the same locations are also obtained using a 3-D FE model shown in Fig. 5.17. The results are plotted in Fig. 5.18 for the two loading conditions. It is seen that the two approaches give almost identical results.

This approach can be extended to the tapered beam shown in Fig. 5.15, for which the width varies continuously with depth. In this case we approximate the width variation by a series of layers of varying width shown in Fig. 5.19. Then a 2-D FE model is developed such that the width W_i at the ith layer is modelled by a proportionally scaled elastic modulus E_i, i.e.,

$$\frac{E_0}{E_i} = \frac{W_0}{W_i}$$

The stress acting on the crack face at the ith layer is also scaled according to

$$f(x) = \frac{f_0\, w(x)}{w_0}$$

As the number of layers increases, the approximation is expected to become more accurate. To study the convergency of the approximation, the strain on the back face of the beam is computed using 5, 10 and 25 material layers for a point load acting on the crack mouth. Figure 5.20 shows the normalized strains for the three configurations. Noticeable oscillation is found in the computed strain when only five layers are used. The oscillation diminishes for 10 layers and disappears almost entirely for 25 layers. The difference between the results for 10 and 25 layers is found to be less than 1.4%.

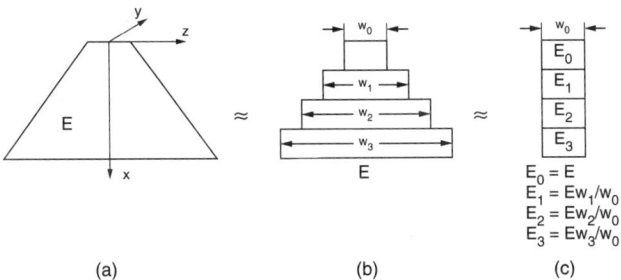

(a)   (b)   (c)

Fig. 5.19. Continuously varying width approximated by a series of layers of varying width.

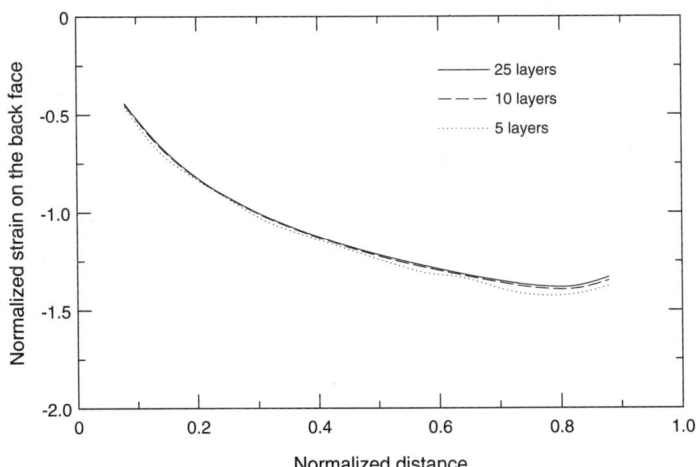

Fig. 5.20. Normalized strain on the back face computed using 5, 10 and 25 material layers.

5.4 Discussion

In this chapter we have explored the capability of the FEM to obtain the compliance functions for a variety of configurations. To further reduce the computing time, one may use a non-evenly spaced element mesh along the plane of the crack. This will be very useful for a time-consuming 3-D FE analysis. However, for most 2-D FE computations the benefit from reduced computing time is not worth the added complexity in setting up such a mesh.

Unlike the analytical approach used in Chapters 3 and 4, for which the computation can be carried out for any given depth of cut, computations using an evenly spaced element mesh are less flexible and compliance·functions are often obtained only for a set of fixed depths. Since the actual increment for a cut of increasing depth varies from one measurement to another, the strains measured in a test usually do not correspond to those computed by FEM. For this reason an interpolation procedure is often required to obtain either the strain at depths for which the compliance functions have been computed or the compliance functions for which the strains are measured.

6

Estimation of Residual Stresses

6.1 Introduction

Before we introduce the methods for estimating residual stresses from measured strains, it is helpful to examine strain responses to some typical residual stresses released by a cut of progressively increasing depth. Two stress distributions are considered, one with the peak stress at one surface (Fig. 6.1.a), another, though unsymmetrical, with the same peak stress on both surfaces (Fig. 6.2.a). Different strain responses are obtained depending upon whether the strain gage is located on the right or left face, as shown in Figs. 6.1.b and 6.1.c, or Figs. 6.2.b and 6.2.c respectively. In general, when a tensile stress is released near the surface, the measured strain is initially negative, and vice versa. This response is expected from the superposition principle demonstrated in Section 2.1 that the measured deformation is the same as that produced by applying the released stress to the faces of the cut with an opposite sign. Alternatively, a more intuitive explanation is that the release of a tensile stress opens the cut and leads to a compressive bending stress on the face opposite the cut. As the depth of cut increases, the measured strain variations will take very different forms. Note the more than tenfold difference in the magnitude of measured peak strains shown in Figs. 6.1.b and 6.1.c. In the first case the high stress near the surface is quickly balanced by a compressive stress below the surface, leading to a very gradual strain variation. In the second case the released stress is seen to result in a small bending stress whose influence increases rapidly as the cut approaching the face with the strain gage. The result shown in Fig. 6.2 is of equal interest: although the peak stress near the surface is the same, the magnitude of the strain differs almost 60%. Clearly, a lower magnitude of the measured strain by no means implies a lower peak stress.

After the strains are measured, the next step is to choose an appropriate method to estimate the residual stress distribution from the measured strains. We will introduce several methods, each of which is characterized by the approximation used for the unknown residual stress, i.e., a continuous function

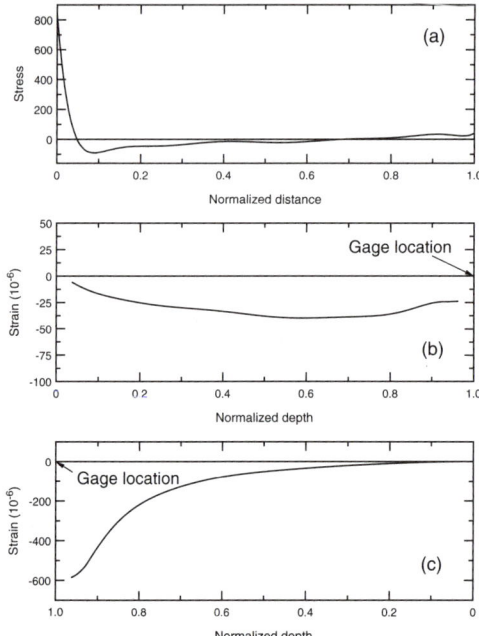

Fig. 6.1. (a) A typical residual stress distribution to be measured by the compliance method; (b) Strain measured when the cut is made from the left to right; (c) Strain measured when the cut is made from the right to left.

defined by a power or polynomial series, a series of strip loads, or a series of piecewise functions. The choice of the method not only dictates how the compliance functions will be expressed but also has a significant influence on the estimated residual stress. Unlike the computation of the compliance functions for which the problem is well defined, the process of estimation has to deal with the uncertainties in the unknown residual stresses as well as various sources of error [119]. For this reason the behavior of a method should be evaluated by experimental validation and numerical simulation or, preferably, an error analysis before it can be used with confidence.

An important part of error analysis is to identify the sources of error associated with the slitting method which may arise from analysis, numerical computation, and/or measurement. The error in analysis comes from the difference between the computed compliance functions and the actual response to the release of residual stresses, and the difference between the stress distribution to be measured and the approximate stress distribution used in the estimation. For example, in a 2-D analysis we usually assume the state of deformation is either in plane stress for a thin specimen or in plane strain for a long specimen. However, many parts have a width that falls between the two extremes. Thus, a deformation assumed to be in either state will differ from the actually measured ones. Also, the residual stresses so far have been assumed to be uniform in the direction along the length of the cut. If this condition is not satisfied, the estimated stress would only be an approximation of the mean stress over the length, and a more accurate estimation would require a three-dimensional analysis. Furthermore, the presence of stress components other than the normal stress as analyzed in Section 2.2, on the faces of cut influences the measured strain, and a computation based on normal stresses alone will be in error, which deserves special attention for near surface measurements.

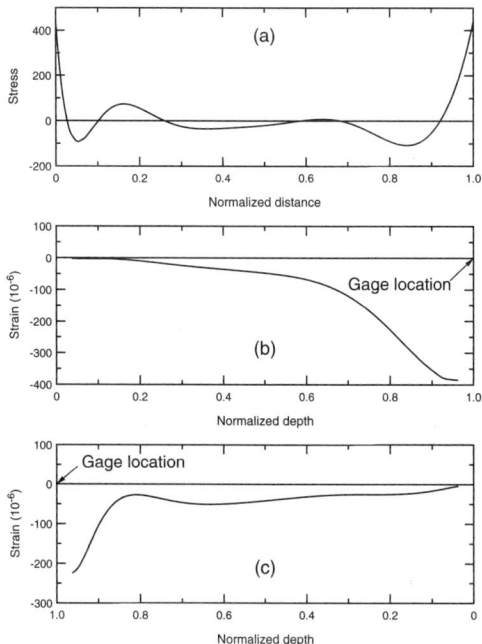

Fig. 6.2. (a) Another residual stress distribution to be measured by the compliance method; (b) Strain measured when the cut is made from the left to right; (c) Strain measured when the cut is made from the right to left.

Since residual stresses are inherently nonuniform and a closed-form solution for a part with a cut of increasing depth is still unattainable, the computation of the compliance functions has to be carried out numerically. For the approach based on LEFM (Chapter 4), in particular, some uncertainties may be present for deep cuts if self-balanced stresses are used in computing the compliances. For the FEM approach (Chapter 5) the discretization of a continuous loading condition on the nodes along the faces of cut could result in a small residue in the resultant force and/or moment, which may have an increasing influence on computed compliance as the depth of cut increases. Furthermore, as will be discussed in this chapter, the properties of the compliance matrix formed by a least squares fit using the compliance functions may become ill-conditioned as the order of estimation increases, making it numerically unstable to estimate higher order stresses.

The measurement error may be decomposed into a continuous varying error and a randomly distributed error. Typical examples of the former are the influences of temperature change, misalignment of the strain gage and weight of the part on the strain measurement. The latter is predominately caused by errors in strain reading and in measurement of the depth of cut. When a cut is made by wire EDM, the depth of cut is usually estimated from the readings of the wire position, which is highly accurate to about 0.0025 mm. However, the small gap between the wire and bottom of the cut changes constantly, and its variation at different depths may be several times of the wire position accuracy. It is possible to estimate the effect of these errors analytically or through numerical simulations [100]. As will be demonstrated in this chapter, when strains are "contaminated" by a randomly distributed error, estimations using a least squares fit usually leads to a more reliable estimation than that using the strain/stress relation directly.

We will start our journey with the approximation using a series of power functions. After examining the strength and weakness of the method, two other approximations that utilize a polynomial series or a series of piecewise functions are then presented. It will be shown that the influence of error on estimated stress is directly related to the stability of the method. The major advantage of the latter two approximations over the first one is the improved stability for estimation of higher order residual stresses.

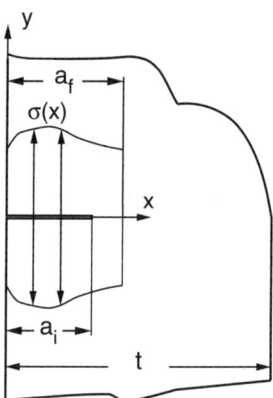

Fig. 6.3. An unknown residual stress distribution $\sigma(x)$ in a part of thickness t released by a progressively increasing cut a_i till the final depth a_f is reached.

6.2 Approximation Using a Power Series

Consider a residual stress distribution $\sigma(x)$ acting on the faces of a cut of increasing depths a_1, ..., a_i, ... a_f shown in Fig. 6.3. When the unknown residual stress is assumed to be continuous, we may approximate it by an nth order power series, i.e.,

$$\sigma(x) = \sum_{i=0}^{n} A_i x^i \tag{6.1}$$

where A_i is the amplitude factor to be determined for the ith order power function. To make our discussion that follows independent of the actual size of the cut or the thickness of the part, the interval defining the distance variable x in Eq. (6.1) is always mapped to a domain [0,1]. Denoting the unmapped distance as x', the mapping leads to $x = x'/a_f$ over the final depth of cut a_f for near surface or partial through-thickness measurement, and $x = x'/t$ over the thickness t for through-thickness measurement. Thus, $\sigma(x)$ is always defined over a domain from 0 to 1.

6.2.1 Least Squares Fit for Stress Estimation

For linear elastic deformation the strain corresponding to the stress given by Eq. (6.1) acting on the faces of a cut of depth a becomes

$$\epsilon(a) = \frac{1}{E'} \sum_{i=0}^{n} A_i C_i(a) \tag{6.2}$$

where $C_i(a)$ is the compliance function for the ith order function in Eq. (6.1) and may be computed using one the approaches presented in Chapters 3-5. To determine the $n+1$ unknowns in Eq. (6.1) or (6.2), we make measurement of strains at $m > n + 1$ depths of cut. If there were no error involved in measurements and computations, and the order of variation of the unknown residual stress were equal or less than that given by Eq. (6.1), identical results would be obtained for the amplitude factors A_i from strains measured at any $n + 1$ depths of cut. In practice, however, some error is always present, and different results will be obtained when strains for a different set of n+1 depths of cut are used. Denoting the strains computed by Eq. (6.2) and those measured by ϵ' and ϵ respectively, the difference between them at the kth depth a_k is given by

$$\Delta\epsilon(a_k) = \epsilon'(a_k) - \epsilon(a_k) = \frac{1}{E'} \sum_{i=0}^{n} A_i C_i(a_k) - \epsilon(a_k) \tag{6.3}$$

Since the error may be negative or positive, we take a summation of the squares of the error over all depths, which from Eq. (6.3) gives

$$\sum_{k=1}^{m} \Delta \epsilon^2(a_k) = \sum_{k=1}^{m} \left[\frac{1}{E'} \sum_{i=0}^{n} A_i C_i(a_k) - \epsilon(a_k) \right]^2 \tag{6.4}$$

Clearly, the minimum of the total error is achieved only if the partial derivatives with respect to A_i for $i = 0$ to n are equal to zero. That is, after changing the summation index of A from i to j, we have

$$\frac{\partial}{\partial A_i} \sum_{k=1}^{m} \Delta \epsilon^2(a_k) = \frac{\partial}{\partial A_i} \sum_{k=1}^{m} \left[\frac{1}{E'} \sum_{j=0}^{n} A_j C_j(a_k) - \epsilon(a_k) \right]^2$$

$$= \sum_{k=1}^{m} \left[\frac{1}{E'} \sum_{j=0}^{n} A_j C_j(a_k) - \epsilon(a_k) \right] C_i(a_k) = 0 \tag{6.5}$$

for $i = 0, ..., n$. In a matrix form Eq. (6.5) becomes

$$\frac{1}{E'} [\overline{C}] \mathbf{A} = \mathbf{B} \tag{6.6}$$

where the vectors

$$\mathbf{A} = [A_0 ... A_i ... A_n]^T, \quad \mathbf{B} = \left[\sum_{k=1}^{m} C_{0k} \epsilon_k ... \sum_{k=1}^{m} C_{ik} \epsilon_k ... \sum_{k=1}^{m} C_{nk} \epsilon_k \right]^T$$

and the compliance matrix

$$[\overline{C}] = \begin{bmatrix} \sum_{k=1}^{m} C_{0k} C_{0k} & \cdots & \sum_{k=1}^{m} C_{0k} C_{jk} & \cdots & \sum_{k=1}^{m} C_{0k} C_{nk} \\ \cdots & \cdots & \cdots & \cdots & \cdots \\ \sum_{k=1}^{m} C_{ik} C_{0k} & \cdots & \sum_{k=1}^{m} C_{ik} C_{jk} & \cdots & \sum_{k=1}^{m} C_{ik} C_{nk} \\ \cdots & \cdots & \cdots & \cdots & \cdots \\ \sum_{k=1}^{m} C_{nk} C_{0k} & \cdots & \sum_{k=1}^{m} C_{nk} C_{jk} & \cdots & \sum_{k=1}^{m} C_{nk} C_{nk} \end{bmatrix}$$

with $C_{ik} = C_i(a_k)$ and $\epsilon_k = \epsilon(a_k)$.

To many readers the derivation from Eqs. (6.3) to (6.6) is more or less known to be the standard procedure for a least squares fit (LSF). Alternatively, Eq. (6.6) can be obtained more directly from the solution of an overdetermined linear system equation. In this case for m measurements of strain we can rewrite Eq. (6.2) in an array form of

$$\frac{1}{E'} [C] \mathbf{A} = \epsilon \tag{6.7}$$

in which $[C]$ is an $m \times (n+1)$ compliance array with elements defined by $C_i(a_k)$ and ϵ is an $m \times 1$ vector of the measured strains $\epsilon(a_k)$. The coefficient vector $\mathbf{A}$ can not be solved directly from Eq. (6.7) because $m > n+1$. However, if we multiply both sides of the equation by the transposed array $[C]^T$, a system of $(n + 1) \times (n + 1)$ equations are obtained, i.e.,

$$\frac{1}{E'}[C]^T[C]\mathbf{A} = [C]^T\boldsymbol{\epsilon} \tag{6.8}$$

It is straightforward to show that $[C]^T[C] = [\overline{C}]$ and $[C]^T\boldsymbol{\epsilon} = \mathbf{B}$. The advantage of expressing the least squares fit by Eq. (6.8) is that the array manipulation can readily be carried out by a computer software such as Mathcad, MATLAB or Mathematica.

6.2.2 Properties of Compliance Matrices

In the previous section we demonstrated how to obtain the compliance matrix $[\overline{C}]$ for residual stress estimation using a least squares fit. Although it is very convenient to have residual stress expressed in terms of a power series, it is known in numerical analysis that a least squares fit using power functions is unsatisfactory because the monomials of x^i for $i > 0$ are very nearly linearly dependent and the matrix becomes rapidly numerically unstable as the order of estimation increases. It is therefore of interest to examine if a similar behavior also exists for the compliance matrix obtained using a power series.

The evaluation uses the condition number of the compliance matrix $cond([\overline{C}])$ which is defined by the product of the matrix norms [1], i.e.,

$$cond([\overline{C}]) = \|[\overline{C}]\|\|[\overline{C}]^{-1}\|$$

Here we use the maximum norm to define the matrix norm,

$$\|[\overline{C}]\| = \underset{1 \leq i \leq n}{Max} \sum_{j=1}^{n} |\bar{c}_{ij}|$$

It can be shown that matrix $[\overline{C}]$ is most stable if the condition number is nearly 1. But if $cond([\overline{C}])$ is very large, say larger then 10^7, then the solution may become unstable for a small perturbation in numerical computation, such as the truncational error due to using single-precision floating-point arithmetic.

For near the surface measurement the compliance matrices are obtained using the body force method for power series of order $n = 0, ..., 6$, and $m = 10$ and 20. It is seen in Fig. 6.4 that $cond([\overline{C}])$ increases very rapidly as n increases. This is true for both $m = 10$ and 20. To further illustrate the impact of the numerical error on higher order compliance matrices for power functions, we calculate the determinants of the compliance matrix $det([\overline{C}])$ using both single and double precision floating-point arithmetic and plot the ratio of the results in Fig. 6.5. As expected, for $n > 4$ the matrix becomes so ill-conditioned that the truncational error produced by a single-precision floating-point computation alone is intolerable for most applications.

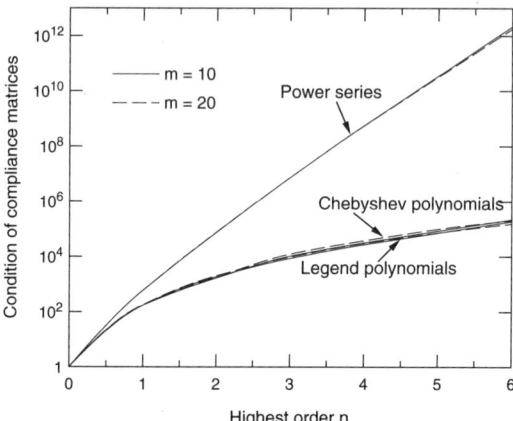

Fig. 6.4. Condition numbers of the compliance matrices computed using power series, Legendre series and Chebyshev series ($n = 0, ..., 6$) for near the surface measurement.

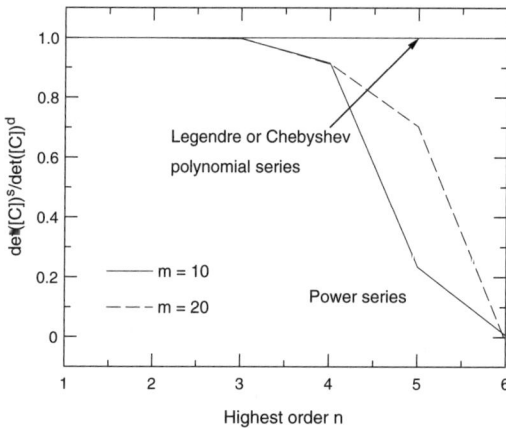

Fig. 6.5. Ratio of the determinants of the compliance matrices for near-the-surface measurement computed by single (superscript s) and double (superscript d) precision floating-point arithmetic ($n = 1, ..., 6$).

6.3 Approximation Using Polynomial Series

To improve the condition of the compliance matrices, Legendre polynomials $L_i(x)$ and Chebyshev polynomials $T_i(x)$, have been used in the place of power functions in Eq. (6.1), which lead to

$$\sigma(x) = \sum_{i=0}^{n} A_i L_i(x) \quad or \quad \sigma(x) = \sum_{i=0}^{n} A_i T_i(x) \tag{6.9}$$

A general form of $L_i(x)$ over a domain $0 \le x \le 1$ has been given in Chapter 4. For Chebyshev polynomials the general form over a domain $0 \le x \le 1$ is given by

$$T_i(x) = cos[i \ cos^{-1}(2x - 1)] \tag{6.10}$$

from which the first five terms are

$$T_0(x) = 1$$
$$T_1(x) = 2x - 1$$
$$T_2(x) = 8x^2 - 8x + 1$$
$$T_3(x) = 32x^3 - 48x^2 + 18x - 1$$
$$T_4(x) = 128x^4 - 256x^3 + 160x^2 - 32x + 1$$
$$T_5(x) = 512x^5 - 1280x^4 + 1120x^3 - 400x^2 + 50x - 1$$

For comparison the condition numbers of $[\overline{C}]$ are also obtained using these two polynomials. It is seen from Fig. 6.4 that $cond([\overline{C}])$ is significantly reduced when either of the polynomials is used. In what follows the Legendre polynomial series and Chebyshev polynomial series will be simply referred to as Legendre series and Chebyshev series.

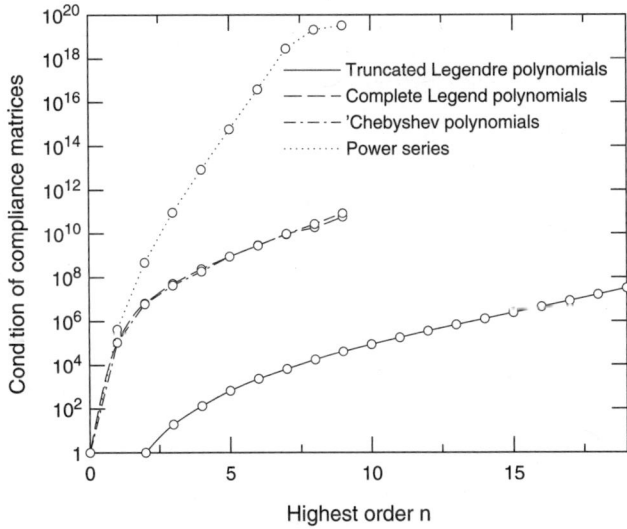

Fig. 6.6. Condition numbers of the compliance matrices computed using power series, Legendre series, Chebyshev series ($n = 0, ..., 9$) and truncated Legendre polynomials ($n = 2, ..., 19$) for through-thickness measurement. x is normalized by thickness and $a_f = 0.96$.

6.3.1 Polynomial Series for Through/Partial-Through-Thickness Measurement

For through or partial-through-thickness measurement the distance x in Eq. (6.9) should be normalized by the thickness or the final depth of cut. The approximation using Legendre series defined over the thickness becomes especially useful because the equilibrium conditions of zero resultant force and moment over the thickness can be satisfied exactly by omitting the first and the second terms in the series. For the same configuration shown in Fig. 4.2, compliance functions are obtained for the approximation using a power series, a Chebyshev series, a Legendre series and a truncated Lengendre series. Figure 6.6 shows the condition numbers of $[\overline{C}]$ for compliance matrices obtained for the four cases plotted against the order of estimation n. It is seen again that the condition number for power functions increases very rapidly as n increases while the truncated Legendre polynomials lead to condition numbers significantly less than those for the complete polynomial series. It is interesting to note that even for $n = 19$ the corresponding condition number is still less than 10^8, which implies that numerical stability is essentially guaranteed for high order approximations using truncated Legendre polynomials as long as the computation is carried out using double precision.

Although the result shown in Fig. 6.6 is very encouraging for the use of truncated Legendre polynomials, two common situations in measurement will limit its use in stress estimation. First, the final depth of cut in computing the compliance matrices for the results shown in Fig. 6.6 is $a_f = 0.963t$. However, some measurements in practice may have to terminate at a depth far less than $0.963t$, and defining the polynomials over the final depth of cut may yield a better conditioned compliance matrix than defining them over the thickness. As a comparison, the condition numbers of the compliance matrices of complete and truncated Legendre series are obtained for different values of a_f and plotted against estimation order n in Fig. 6.7. It is seen that the condition of the compliance matrix for truncated Legendre polynomials deteriorates rapidly as a_f/t decreases while its condition for complete series is much less influenced and actually improves as a_f/t decreases.

The second issue arises when the error in measured strain which increases with depth is in such a form that may be simulated by an unbalanced load. In this case the use of a complete polynomial series allows the error to be absorbed by the unbalanced terms and therefore reduces its influence on the overall residual stress estimation. To illustrate this let us recall that in Chapter 5 a LEFM solution is found to differ increasingly with a FEM solutions for $a/t > 0.7$. For a residual stress distribution given by

$$\sigma(x) = 100L_2(x) + 100L_3(x) - 50L_4(x)$$

the change of strain on the surface opposite the cut for a through-thickness measurement may be computed using the compliances obtained by FEM or

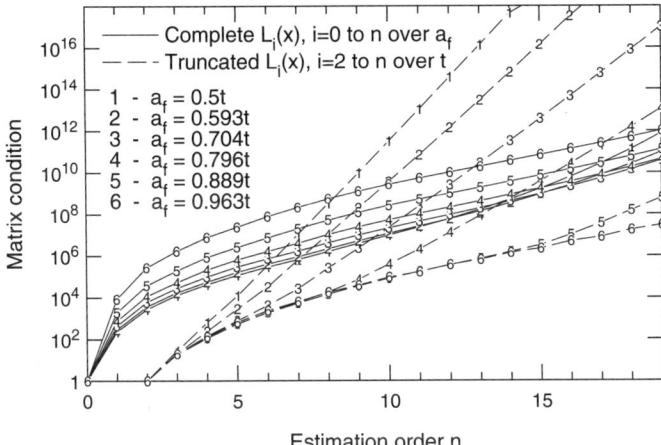

Fig. 6.7. Condition numbers of the compliance matrices for through-thickness measurement computed using Legendre series ($n = 0, 1..., 19$) with x normalized by a_f and truncated Legendre polynomials ($n = 2, ..., 19$) with x normalized by thickness.

by LEFM solution, Eq. (4.10), as shown in Fig. 6.8. Taking the strain computed by FEM (solid line) as the "measured" strain, the residual stress is first estimated using the LEFM compliances for a truncated Legendre series, and the result is shown in Fig. 6.9. Next, for the same strain the stress is estimated using the LEFM compliances for a complete Chebyshev series, and the result is then converted to a residual stress by a least squares fit using a truncated Legendre series. Note that the last step is necessary because the stress obtained by Chebyshev polynomials does not satisfy the equilibrium conditions and that a least squares fit over the obtained stress using truncated Legendre polynomials is numerically more stable than that over the strains. Figure 6.9 shows a comparison of the two approaches: a complete Chebyshev series leading to a result very close to the original stress while a truncated Legendre series yielding a considerable deviation for $a/t < 0.5$.

In summary, an approximation using a complete polynomial series is suitable for both partial-through and through-thickness measurements. When a Chebyshev series is used, the equilibrium conditions can be imposed by a least squares fit over the estimated stress using a truncated Legendre series defined through the thickness. The use of a truncated Legendre series yields a superior compliance matrix condition, which is useful for high order approximations. However, this condition holds true only for through-thickness measurement ($a/t > 0.9$). Also, the estimation is shown to be more susceptible to errors that resemble strain produced by an unbalanced load than that using a complete polynomial series. As will be shown in the next section, the influence of this type of error can be reduced by using a weighted least squares fit.

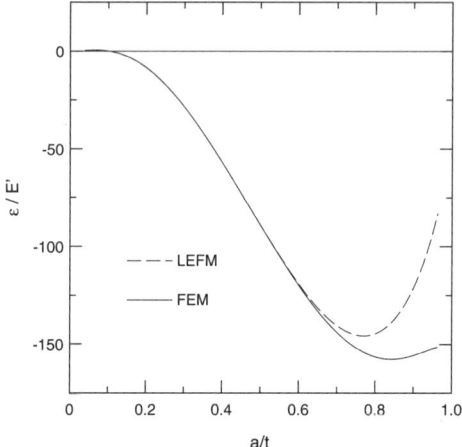

Fig. 6.8. Strains computed using FEM and Eq. (4.10) for $\sigma_4(x)$

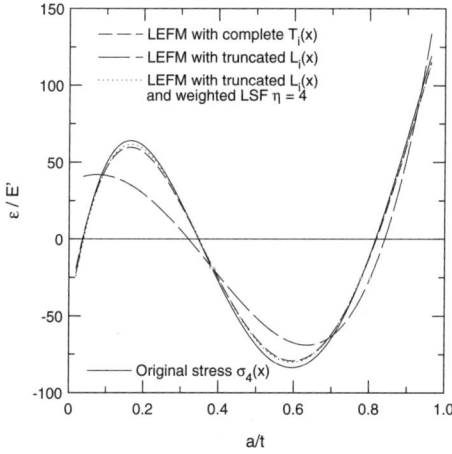

Fig. 6.9. Comparison of $\sigma_4(x)$ to stresses estimated using different compliances obtained by LEFM, Eq. (4.10), for a complete Chebyshev series and a truncated Legendre series. Also shown are the results using weighted least squares fit (LSF).

6.3.2 Weighted Least Squares Fit for Through-Thickness Measurement

A feature of least squares fit (LSF) is that its result is more influenced by data with larger magnitude, which is desirable if the process is applied directly to the data of variables to be estimated. For residual stress measurements, however, the fitting is over a strain variation that increases as the depth of measurement increases, yielding a result that is much more dominated by the data in the last 40% of the measurement. It is evident from Fig. 6.9 that for the truncated Legendre series the agreement of the stress estimated for last 40% depth is much better than that for the first 40%. To compensate this problem, a weighted least squares fit is often used to apply a more desirable influence of data on the result. Turning to the equation array in Eq. (6.7), we multiply each equation by a weight w_k on both sides of the equation and in an array form we have

$$[w][C]\mathbf{A} = E'[w]\epsilon \qquad (6.11)$$

where $[w]$ is an $m \times m$ diagonal matrix with elements $w_{kk} = \sqrt{w(a_k/t)}$ varying with depth a_k. The matrix form of a weighted least squares fit is thus given by

$$\frac{1}{E'}[C]^T[W][C]\mathbf{A} = [C]^T[W]\epsilon \qquad (6.12)$$

where the weight matrix $[W]$ is an $m \times m$ diagonal matrix with elements $W_{kk} = w(a_k/t)$. The choice of $w(a_k/t)$ is dependent not only on the variation of the compliance with a/t but also on the distribution of error in measured strains. Typically, the strain measured in the first few cuts (up to 25% of the thickness) is fairly small, and the influence of strain reading truncation and offset is much more pronounced than that in the rest of measurement. When both factors are taking into account, $w(a_k/t)$ may take the following form,

$$w(a_k/t) = [(1 - a_k/t)/\sqrt{a_k/t}]^\zeta \qquad (6.13)$$

where the exponent ζ is a positive real number. Utilizing the same example in the last section, the residual stress is again estimated by a weighted LSF with different values of ζ for compliance matrix computed for a truncated Legendre series. A comparison of the estimated coefficients are tabulated below.

$\sigma(x)$		LSF ($\zeta = 0$)		Weighted LSF, $L_{i \geq 2}$					
Coefs.	L_i from T_{0-4}	$L_{i \geq 0}$	$L_{i \geq 2}$	$\zeta = 1$	$\zeta = 2$	$\zeta = 3$	$\zeta = 4$	$\zeta = 5$
$A_2 = 100$	94.01		94.03	106.81	101.23	100.27	99.73	98.54	95.28
$A_3 = 100$	95.22		95.32	78.95	102.86	106.80	105.59	100.42	87.65
$A_4 = -50$	-46.24		-46.20	7.30	-27.32	-43.27	-53.56	-65.44	-82.76

in which the second column shows the coefficients for a truncated Legendre series obtained by a LSF over the stress given by a Chebyshev series that was estimated over the strain variation. It is seen that for $2 < \zeta < 3$ the weighted LSF greatly improved the estimation. The result represents an "idealized" application of the weighted least squares fit in which the data for small a/t are only influenced by the round-off error from strain readings, and for large a/t the error from compliance computation becomes increasingly significant.

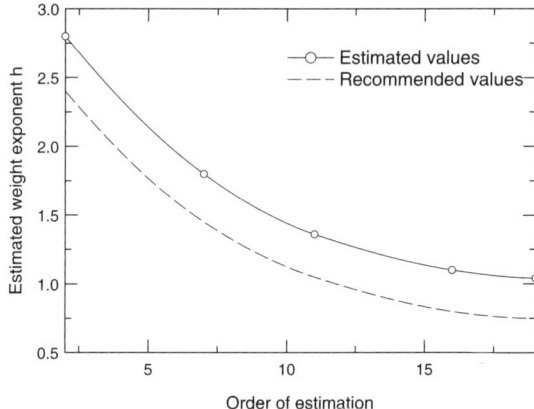

Fig. 6.10. Estimated values of the exponent in weight function as a function of the order of approximation.

Carrying out similar estimations for higher order stresses, we found that the optimum value of ζ decreases with the increase of approximation orders, as shown in Fig. 6.10. In practice an uncertainty larger than the rounding-off error often exists for strain measured at $a/t < 0.2$, a reduced value of ζ should be used, as suggested in the dashed line in Fig. 6.10. Also a weighted LSF improves the estimation mostly in the region of $a/t < 0.2$ and may worsen the result for $a/t > 0.9$. In that case it can be turned off by setting ζ to zero if one wants to minimize the influence of error that occurred at $a/t < 0.2$ on the stress estimated for $a/t > 0/9$.

So far we have assumed that the residual stresses to be estimated can be sufficiently approximated by a series of polynomials of limited order. In many engineering applications, however, the residual stresses are often caused by abrupt changes in material properties and geometries, or by localized incompatible deformation, such as shot-peening. In these cases the stress gradient may become very steep or even discontinuous, and the estimation using a continuous stress will require a very high order of approximation that becomes unstable in estimation. As an attempt to solve this problem, the stress is usually approximated by a series of piecewise functions.

6.4 Approximation Using Strip-Loads

Consider a residual stress distribution acting on the faces of a cut of increasing depths $a_1, ..., a_i, ... a_n$ shown in Fig. 6.11-a. From linear superposition the stress distribution can be replaced by a series of strip loading over each increment of the cut, denoted by $\sigma(x_i)$ with $a_{i-1} \leq x_i \leq a_i$ and $i = 1, ...n$. Since the variation of stress in each increment is unknown, we may approximate $\sigma(x_i)$ by a series of uniform stress σ_i shown in Fig. 6.11-b with a magnitude to be determined. As a further simplification the uniform stresses can also be approximated by a series of point loads f_i acting at the mid-point of each increment shown in Fig. 6.11-c. The magnitude of the uniform strip loading σ_i is then related to the point load f_i by

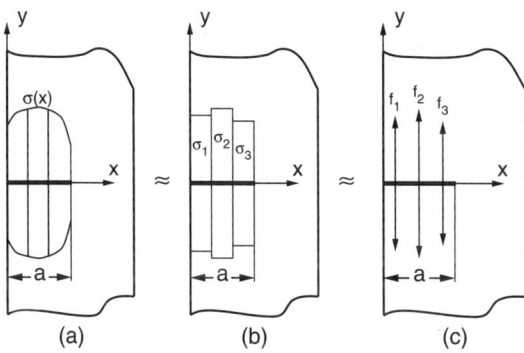

(a)   (b)   (c)

Fig. 6.11. (a) An unknown residual stress distribution on cut faces is approximated by a series of uniform strip loading (b), which is further approximated by a series of point loads (c).

$$\sigma(x_i) \approx \sigma_i = \frac{f_i}{\Delta a_i}$$

where $\Delta a_i = a_i - a_{i-1}$. From the strain computed by applying the jth uniform stress or point load on the faces of the cut of depth a_i, the corresponding compliance functions can be expressed as

$$C_{ij} = \frac{\Delta a_j E'}{f_j} \epsilon_{ij} = \frac{E'}{\sigma_j} \epsilon_{ij} \qquad for \ \ j \le i$$

from which the strain ϵ_i for the ith cut subjected to stresses from σ_1 to σ_i can be expressed as

$$\epsilon_i = \sum_{j=1}^{i} \epsilon_{ij} = \sum_{j=1}^{i} \sigma_j \frac{C_{ij}}{E'} \tag{6.14}$$

For $i = 1$ to n Eq. (6.14) the compliance matrix for strip loads becomes

$$\frac{1}{E'}[\overline{C}] \cdot \sigma = \epsilon \tag{6.15}$$

where

$$[\overline{C}] = \begin{bmatrix} c_{11} & 0 & 0 & 0 & 0 \\ \cdots & \cdots & 0 & 0 & 0 \\ c_{i1} & \cdots & c_{ij} & 0 & 0 \\ \cdots & \cdots & \cdots & \cdots & 0 \\ c_{n1} & \cdots & c_{nj} & \cdots & c_{nn} \end{bmatrix}$$

and

$$\boldsymbol{\sigma} = [\sigma_1 ... \sigma_j ... \sigma_n]^T \qquad \boldsymbol{\epsilon} = [\epsilon_1 ... \epsilon_j ... \epsilon_n]^T$$

in which a superscript T is used to denote a row vector transposed from a column vector.

When the strain vector on the left side of Eq. (6.15) is replaced by a set of measured strains, the unknown vector $\boldsymbol{\sigma}$ can be solved readily by a forward substitution. That is,

$$\sigma_1 = E' \frac{\epsilon_1}{C_{11}}$$

$$\sigma_2 = \frac{E'}{C_{22}} \left(\epsilon_2 - \frac{\sigma_1}{E'} C_{21} \right)$$

$$\cdots$$

$$\sigma_n = \frac{E'}{C_{nn}} \left(\epsilon_n - \sum_{j=1}^{n-1} \frac{\sigma_j}{E'} C_{nj} \right) \qquad (6.16)$$

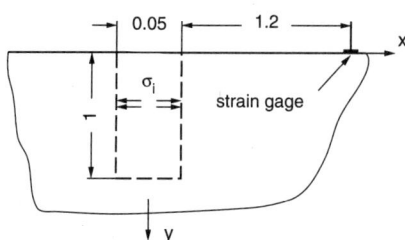

Fig. 6.12. A strain gage located near the cut for near surface residual stress measurement.

6.4.1 Error Analysis

Assuming that the compliance functions can be computed accurately, the error in estimated residual stresses is then associated directly with the error in measured strains, which for linear deformation corresponds to those measured for a stress-free body. Denoting the error in measured strains by $\Delta\epsilon$ and the corresponding error in stresses by $\Delta\sigma$ respectively, the error in the stresses estimated from Eq. (6.16) may be expressed as

$$\Delta\sigma_1 = \frac{E'}{C_{11}}\Delta\epsilon_1$$

$$\Delta\sigma_2 = \frac{E'}{C_{22}}(\Delta\epsilon_2 - \frac{\Delta\sigma_1}{E'}C_{21}) = \frac{E'}{C_{22}}(\Delta\epsilon_2 - \Delta\epsilon_1\frac{C_{21}}{C_{11}})$$

$$\Delta\sigma_3 = \frac{E'}{C_{33}}[\Delta\epsilon_3 - (\frac{\Delta\sigma_1}{E'}C_{31} + \frac{\Delta\sigma_2}{E'}C_{32})]$$

$$= \frac{E'}{C_{33}}[\Delta\epsilon_3 - \Delta\epsilon_1(\frac{C_{31}}{C_{11}} - \frac{C_{32}}{C_{11}}\frac{C_{21}}{C_{22}}) - \Delta\epsilon_2\frac{C_{32}}{C_{22}}]$$

$$\cdots$$

$$\Delta\sigma_n = \frac{E'}{C_{nn}}(\Delta\epsilon_n - \sum_{j=1}^{n-1}\frac{\Delta\sigma_j}{E'}C_{nj}) = \frac{E'}{C_{nn}}(\Delta\epsilon_n + \sum_{j=1}^{n-1}\Delta\epsilon_j C_{nj}^r) \qquad (6.17)$$

where the accumulated compliance C_{nj}^r for $n \geq 2$ can be computed recursively by

$$C_{21}^r = -\frac{C_{21}}{C_{11}}, \quad \cdots, \quad C_{ij}^r = -\sum_{k=j}^{i-1}\frac{C_{ik}}{C_{kk}}C_{kj}^r \quad \left(\begin{array}{l}j = 1, 2, ..., n-1 \\ i = j+1, ..., n\end{array}\right)$$

Equation (6.17) indicates that the error in the stress estimated at depth a_n is affected by the error in the strain measured at that depth as well as those in the previous depths. Clearly, the propagation of the error from a previous step, say step j, to the present step n, depends on the ratio of the accumulated compliance C_{nj}^r to the compliance C_{nn}.

To illustrate the behavior of the error propagation for a unit perturbance, say one micro-strain, at a single interval, we first consider a numerical example for measurement of near surface residual stresses. In this case the location of strain measurement is on the surface near the cut. Using the normalized dimension shown in Fig. 6.12, the compliances for strip loads consisting of ten or twenty intervals are computed for $E' = 30 \times 10^6$ psi by the body force method described in Chapter 3. From Eq.(6.17) the stress responses are obtained and displayed in Fig. 6.13-a. It is seen that except for a reversed overshoot at the most adjacent interval, the influence of perturbance decreases rapidly and becomes almost negligibly small when the distance is larger than 3 intervals. Also, the magnitude of the maximum error increases considerably as the size of the load interval decreases.

Another kind of error of interest corresponds to a static shift which occurs, for example, at the first interval and remains constant over the rest of the intervals. For the same configuration, the influence of the error is obtained and, as shown in Fig. 6.13-b, is marked by a very rapid decay, which vanishes almost

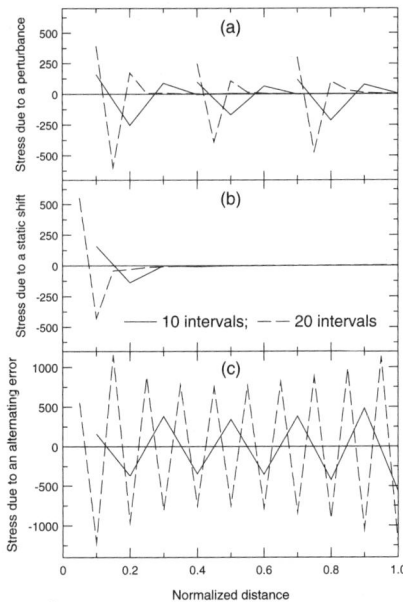

Fig. 6.13. Error for near-surface residual stress measurement due to (a) a unit perturbance applied in a single interval, (b) a static shift over all intervals and (c) distributed errors with a unit amplitude and alternating sign.

entirely in two intervals. The very localized influence is produced by error cancellation from the superposition of the responses to all the perturbances. This implies that if an error is distributed with a sign alternating from one interval to another, as found in most randomly distributed errors, the resulting error in stresses will exhibit a significant amount of scatter. Figure 6.13-c shows a typical response to a distributed error with a unit magnitude and alternating sign.

We now turn to the measurement of through-thickness residual stresses for which the location of the strain measurement is on the back face opposite the cut, as shown in Fig. 4.1. Strip loads consisting of ten or twenty intervals are used to compute the influence of the error due to a single perturbation (i.e., one micro-strain) at an interval, a static shift over all intervals or an alternating strain of unit magnitude. For $E' = 30 \times 10^6$ psi the responses to the three kinds of error in strain measurements are estimated and shown in Figs. 6.14-a, 6.14-b and 6.14-c respectively. It is seen that the variation of the error in estimated stress is similar to those found for near surface measurement but the magnitude is considerably lower. One distinct feature of the through-thickness measurement is that the response to an alternating error of constant magnitude decreases steadily with the increase of the depth of the cut.

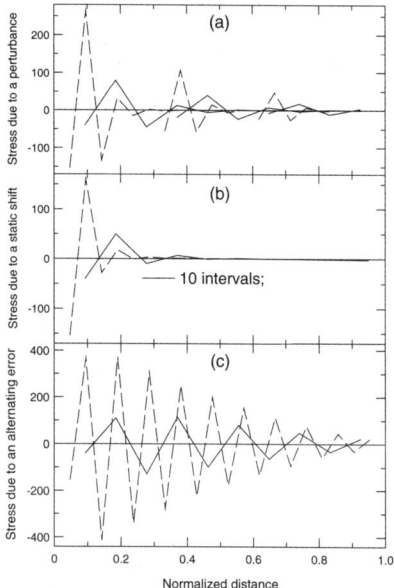

Fig. 6.14. Error for through thickness residual stress measurement due to (a) a unit perturbance applied in a single interval, (b) a static shift over all intervals and (c) distributed errors with a unit amplitude and alternating sign.

6.4.2 Discussion

The approximation using strip-loads has its root from the layer removal method which assumes that the stress in each layer being removed is uniform. It has also been used in the hole drilling method for measurement of non-uniform residual stresses. In fact, the analysis obtained for the near surface residual stress measurement applies equally to the hole drilling method when the compliance is replaced by that for a hole of increasing depth.

The error analysis presented in this section demonstrates that when a single perturbance or a static shift is involved, the approximation exhibits a very good resistance to error propagation. But when a randomly distributed error is present, a significant scatter will be introduced in estimated stresses. Also, when the magnitude of the error is independent of the size of the interval and depth of cut, the influence of the error on the estimated stress increases inversely with the size of the interval.

For through-thickness measurement the approximation is found to be less influenced by the error in measured strains than that for near-the-surface measurement. Also, the influence of a randomly distributed error is found to decrease steadily with the depth of cut.

The main feature of the strip-load approximation is that it requires no explicit assumption for the residual stress distribution and the resolution of the approximation may be improved by reducing the size of intervals. However, as was shown earlier, the reduction of interval size will significantly increase the influence of error in measured strains on estimated stresses. To obtain a stable estimation it is often necessary to omit a large part of the measured data. To make full use of the data and improve the estimation of the stress in each interval, the strip-loads approach is extended in the next section.

6.5 Overlapping-Piecewise Functions for Near-Surface Measurement

When residual stress is released by cutting using wire EDM, a large number of data can be readily obtained in a single measurement. To make use of the full set of data a more versatile approximation that uses overlapping piecewise functions is developed in [60]. In this case the function defined in a sub-interval can be either a uniform, linear, or quadratic polynomial. Also, the number of data points in each sub-interval is always larger than that of the coefficients in the polynomial. Thus, the coefficients can be obtained by using a least squares fit. Since an approximation based on polynomials tends to give a better estimation near the center of the interval than at ends, an overlap is specified by sharing two data points between two adjacent sub-intervals. This ensures that the tail portion of the interval will not be used for calculation of the compliance for the subsequent depths of the cut. As shown in Fig. 6.15, by a numerical simulation, when the last data point in a sub-interval is removed by overlapping, an improved estimation of the original stress distribution is obtained. This procedure is necessary because directly imposing continuity at the end points between two adjacent intervals will not lead to a better approximation nor allow the application of a conventional least squares fit. After the piecewise functions have been obtained, we delete the last points of all intervals except the last one. The value of the stress at the boundary of two adjacent intervals is determined by the average of the stresses calculated from the two coinciding points.

Based on the above discussion, an overlapping piecewise function for the compliance method may be defined as

$$\sigma_y(x) = \sum_{i=0}^{n_j} \sigma_i^j L_i^j \left(\frac{x - x_{aj}}{x_{bj} - x_{aj}} \right) = \sum_{i=0}^{n_j} \sigma_i^j L_i^j(s) \quad j = 1, ..., N \tag{6.18}$$

in which σ_{ij} is the stress amplitude for the ith order polynomials $L_{ij}(s)$ defined by the normalized local distance $0 \leq s \leq 1$ in the jth sub-interval with end points x_{aj} and x_{bj}. The subscripts aj and bj are integer indices denoting the sequence of the global data points. If there are m_j data points in the jth sub-interval, the indices of the data points can be expressed as

$$a_j = 1 + \sum_{k=1}^{j-1} m_k - 2(j-1); \quad bj = \sum_{k=1}^{j} m_k - 2(j-1)$$

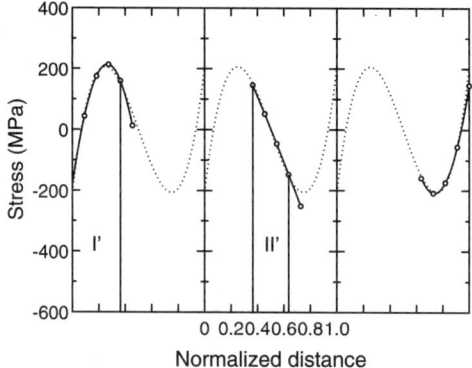

Fig. 6.15. Illustration of the procedure for estimation of residual stress using overlapping piecewise functions. The strain used for the simulation are given in Table 6.1. A quadratic function was used in each sub-interval.

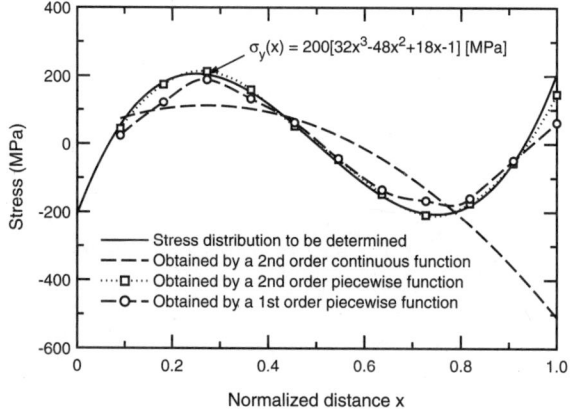

Fig. 6.16. Stress distributions to be obtained (solid line) and that obtained by a quadratic continuous function (dashed line). A third order continuous function would provide an exact prediction. The stress distributions obtained by using linear (dashed data line) and quadratic (solid data line) overlapping piecewise functions are also shown.

To illustrate the use of a continuous function and piecewise functions in residual stress estimation, we consider the same numerical example as shown in Figs. 6.15 and 6.16, in which the residual stress distribution is given by a third order polynomial. We assume that a strain gage of length 0.813 mm (0.032 inch) is installed on the surface and that a cut of width 0.1 mm (0.004 inch) is introduced with the change of strain being measured at eleven depths. The edge of the cut is located 2.16 mm (0.085 inch) from the center of the strain gage. For $E = 206$ GPa and $\nu = 0.3$, the strains due to introducing

Table 6.1. Computed strain to be used in simulation of residual stress measurement for Figs. 6.15 and 6.16

Depth of cut		Strain measured
(in)	(mm)	(10^{-6})
0.007	0.1778	7.4
0.014	0.3556	-0.9
0.021	0.5334	-36.8
0.028	0.7112	-90.6
0.035	0.8890	-144.7
0.042	1.0668	-183.7
0.049	1.2446	-201.1
0.056	1.4224	-199.3
0.063	1.6002	-186.0
0.070	1.7780	-172.4
0.077	1.9558	-168.9

the cut are computed and tabulated in Table 6.1. The compliance functions are tabulated in Table 6.2 for an estimation based on Legendre series up to the third order. Using a conventional least squares fit, the three coefficients for a second order approximation are obtained, and the estimated stress distribution is shown in Fig. 6.16 as a dashed line. As expected, the result is not very satisfactory because the variation of a third order stress field simply cannot be described by any second order stress field. However, if a third order term were included in the estimation, the prediction would have been a perfect agreement. This indicates that the approximation based on a continuous function may lead to a serious error unless the variation of the stress field is equal to or less than that of the function used in approximation. We now turn to the piecewise least squares fit to estimate a higher order stress field with a lower order stress function. The eleven data points are divided into three sub-intervals, as shown in Table 6.3. A second order stress function with three unknowns is specified in each interval using the normalized distance s. Thus, all functions are defined by a normalized distance x between zero and one. The compliance functions are calculated by applying each term of the stress function to each sub-interval. The results are tabulated in Table 6.3. The compliance functions due to stress fields acting on the non-overlapping portion of each previous sub-interval (I' and II' in Fig. 6.15) are also computed for the depths of the cut in the current interval and are tabulated in Table 6.4.

Table 6.2. Compliance functions for Legendre polynomials

Depth (mm)	uniform	linear	quadratic	cubic
0.1778	2.070847e-02	1.907742e-02	1.608570e-02	1.221652e-02
0.3556	7.306154e-02	6.186869e-02	4.311957e-02	2.256442e-02
0.5334	1.504123e-01	1.165467e-01	6.507529e-02	1.842168e-02
0.7112	2.450979e-01	1.729382e-01	7.417539e-02	2.599936e-03
0.8890	3.487803e-01	2.230601e-01	6.939801e-02	-1.571153e-02
1.0668	4.536640e-01	2.618972e-01	5.447387e-02	-2.797554e-02
1.2446	5.534278e-01	2.874390e-01	3.529676e-02	-3.033694e-02
1.4224	6.465609e-01	3.007396e-01	1.729695e-02	-2.407808e-02
1.6002	7.282980e-01	3.027268e-01	5.108289e-03	-1.425016e-02
1.7780	7.991397e-01	2.960605e-01	1.154240e-03	-7.396426e-03
1.9558	8.591951e-01	2.831096e-01	6.378136e-03	-9.856569e-03

In the following step-by-step discussion the compliance functions tabulated in Tables 6.3 and 6.4 will be used to estimate the residual stress from the strains given by the second column in Table 6.1.

1 The values of the compliance functions for sub-interval I given in Table 6.3 are used to estimate the quadratic function from strains given in Table 6.1 by using a least squares fit over the entire sub-interval. This leads to

$$\sigma_1(x) = \sigma_{01} + \sigma_{11}s + \sigma_{21}s^2 = -179.61 + 1346.6s - 1153.9s^2 \ [MPa]$$

in which the second subscript is used to denote the stress function for the first sub-interval.

2 The compliance functions given in Table 6.4 for the subsequent two sub-intervals due to the stress acting on the non-overlapping portion of sub-interval I are used to obtain the corresponding strains which are then subtracted from the original strains given in Table 6.1 starting from the beginning of the second sub-interval.

3 The stress field for the second sub-interval is obtained by using the same procedure as that in Step 1 from the compliances given in Table 6.3 and the strain modified in Step 2. This leads to

$$\sigma_2(x) = \sigma_{02} + \sigma_{12}s + \sigma_{22}s^2 = 236.35 - 443.41s - 44.649s^2 \ [MPa]$$

4 The same as Step 2 but the strains for the last sub-interval are obtained using the compliances given in Table 6.4 and the strain modified in Step 2.

Table 6.3. Compliance functions for 2nd order piecewise functions over each sub-interval

Depth (mm)	uniform	linear	quadratic
	Sub-interval I		
0.1778	2.070847e-02	1.794151e-03	2.180830e-04
0.3556	7.306154e-02	1.231213e-02	2.933353e-03
0.5334	1.504123e-01	3.725217e-02	1.311560e-02
0.7112	2.450979e-01	7.937575e-02	3.674728e-02
0.8890	3.487803e-01	1.382922e-01	7.887057e-02
	Sub-interval II		
0.7112	3.244840e-02	2.758984e-03	3.332675e-04
0.8890	8.609175e-02	1.407493e-02	3.305859e-03
1.0668	1.502928e-01	3.575419e-02	1.233642e-02
1.2446	2.177859e-01	6.701878e-02	3.020483e-02
1.4224	2.850388e-01	1.065379e-01	5.890979e-02
	Sub-interval III		
1.2446	2.541632e-02	2.163546e-03	2.628859e-04
1.4224	6.218629e-02	1.002889e-02	2.347220e-03
1.6020	1.018400e-01	2.362804e-02	8.062372e-03
1.7780	1.408904e-01	4.197721e-02	1.863072e-02
1.9558	1.772019e-01	6.363161e-02	3.447400e-02

5 Using the compliances given in Table 6.3, the stress function in the third sub-interval is obtained

$$\sigma_3(x) = \sigma_{03} + \sigma_{13}s + \sigma_{23}s^2 = -25.563 - 875.77s + 1046.9s^2 \ [MPa]$$

The estimated stress distribution, as shown in Fig. 6.16, is in very close agreement with the original one. The improvement over the estimation based on a continuous second order function is very significant. As a comparison, a computation based on a first order piecewise least squares fit over five sub-intervals is carried out, and the estimated stress distribution is also shown in Fig. 6.16. The result is still more satisfactory than that using a single continuous second order function.

Table 6.4. Compliance functions for 2nd order piecewise functions over first portion of each sub-interval

Depth (mm)	uniform	linear	quadratic
	Sub-interval I		
0.7112	2.126495e-01	9.524621e-02	5.950503e-02
0.8890	2.626885e-01	1.209371e-01	7.689378e-02
1.0668	3.033712e-01	1.416892e-01	9.085553e-02
1.2446	3.356419e-01	1.581624e-01	1.019210e-01
1.4224	3.615220e-01	1.714038e-01	1.108162e-01
1.6020	3.810824e-01	1.814887e-01	1.176053e-01
1.7780	3.958520e-01	1.891674e-01	1.227853e-01
1.9558	4.067581e-01	1.949050e-01	1.266680e-01
	Sub-interval II		
1.2446	1.923696e-01	8.267574e-02	5.054392e-02
1.4224	2.228525e-01	9.866199e-02	6.150232e-02
1.6020	2.453756e-01	1.102895e-01	6.939558e-02
1.7780	2.623973e-01	1.190043e-01	7.528094e-02
1.9558	2.752350e-01	1.255602e-01	7.970287e-02

6.5.1 Influence of Experimental Error on Overlapping-Piecewise Functions for Near-Surface Measurement

Consider a hypothetical stress distribution given by a fourth order Chebyshev polynomial with a magnitude of 310 MPa at the surface as shown in Fig. 6.17. Both linear and quadratic piecewise functions are used to estimate the stress field. The configuration of the cut and strain measurements is shown in Fig. 6.18. The values of elastic modulus and Poisson's ratio are taken to be 205 GPa and 0.3 respectively. The "exact" strain due to the release of the stresses can be calculated by applying the same stresses on the faces of the cut. Supposing that a cut is made with 25 increments during cutting, the strains calculated as a function of cutting depth are shown in Table 6.5.

Estimations based on the "exact strain" are carried out using linear and quadratic piecewise functions. Eight sub-intervals are used for the linear piecewise functions and six sub-intervals for the quadratic piecewise functions. The number of data points is the same for all the sub-intervals except the last one. The estimated stress distributions are also shown in Fig. 6.17 as a dashed

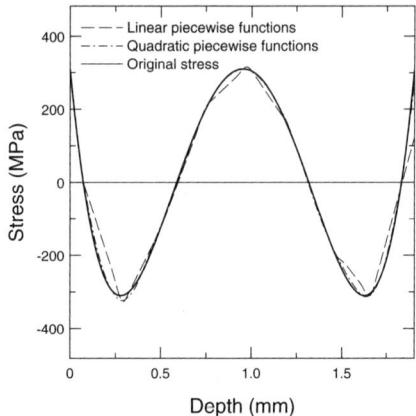

Fig. 6.17. A stress distribution given by a fourth order Chebyshev polynomial used as the original stress field for numerical evaluation. Estimations using linear and quadratic piecewise functions are also shown.

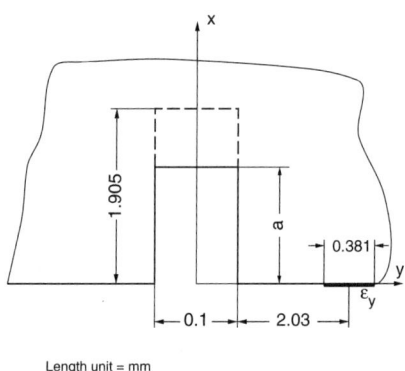

Length unit = mm

Fig. 6.18. The configuration of the cut and strain gage used in numerical simulation.

line for the linear piecewise functions and as a dotted line for the quadratic piecewise functions. It is seen that the second order piecewise functions give an almost exact estimation of the fourth order stress field while the first order piecewise functions are also satisfactory in overall estimation.

Supposing that a zero shift with a magnitude of 6 $\mu\epsilon$ is present, the "measured strains" are shifted by the same amount from the "exact strains." The corresponding stress distribution is estimated by using the same linear and quadratic piecewise functions and the results are shown in Fig. 6.19. It is seen that the influence of zero shift on stress estimations is very small except for the first sub-interval.

We now numerically generate a random error ϵ_r with a magnitude varying from 1 $\mu\epsilon$ to 4 $\mu\epsilon$ at each data point. The error is then superimposed on

Table 6.5. "Exact strains" to be used in simulation of measurement of the residual stress shown in Fig. 6.17

Depth		Strain	Depth		Strain
in	mm	$(\mu\epsilon)$	in	mm	$(\mu\epsilon)$
0.003	0.0762	-3.6	0.042	1.0668	135.3
0.006	0.1524	-3.4	0.045	1.1430	115.4
0.009	0.2286	6.7	0.048	1.2192	97.1
0.012	0.3048	27.4	0.051	1.2954	82.2
0.015	0.3810	56.3	0.054	1.3716	72.9
0.018	0.4572	89.0	0.057	1.4478	69.3
0.021	0.5334	121.1	0.060	1.5240	71.5
0.024	0.6096	148.3	0.063	1.6002	78.2
0.027	0.6858	167.6	0.066	1.6764	87.9
0.030	0.7620	177.5	0.069	1.7526	98.0
0.033	0.8382	177.6	0.072	1.8288	105.7
0.036	0.9144	169.2	0.075	1.9050	107.9
0.039	0.9906	154.3			

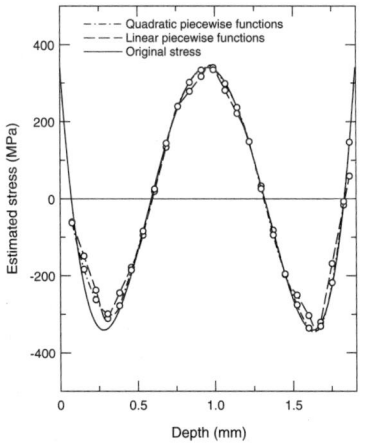

Fig. 6.19. Estimation of the fourth order stress distribution using linear and quadratic piecewise functions from strains subjected to a zero shift of 6 $\mu\epsilon$. The data points correspond to the depths where strain are calculated.

Table 6.6. Five sets of random strain errors used in numerical simulation

Set No.	1	2	3	4	5	No. Set No.	1	2	3	4	5
Depth No.			$\mu\epsilon$			Depth No.			$\mu\epsilon$		
1	2	-3	-3	2	-1	14	2	2	-1	4	-3
2	-1	2	4	-3	2	15	4	4	2	-3	-3
3	-1	-1	2	2	2	16	2	-3	-3	-3	4
4	-3	-1	4	-3	-3	17	2	-1	-3	-3	2
5	-1	-1	-1	-3	2	18	-3	-1	-3	2	-1
6	-3	2	2	4	-3	19	4	4	-3	-3	4
7	4	-1	-1	-3	-1	20	-1	-3	-3	-1	4
8	2	4	-1	2	-3	21	-3	2	2	2	-3
9	2	-1	-3	4	2	22	4	-1	2	2	2
10	2	-3	2	-1	-1	23	4	-3	-3	-1	-3
11	2	-3	2	-3	4	24	-3	2	-1	2	-3
12	-3	2	-3	-3	-1	25	2	2	-3	2	-1
13	-1	-3	-1	-3	4						

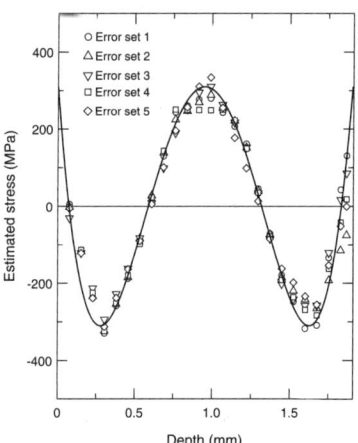

Fig. 6.20. Estimation of the fourth order stress distribution using linear piecewise functions from strains subjected to different random errors. The data points correspond to the depths where strain are calculated.

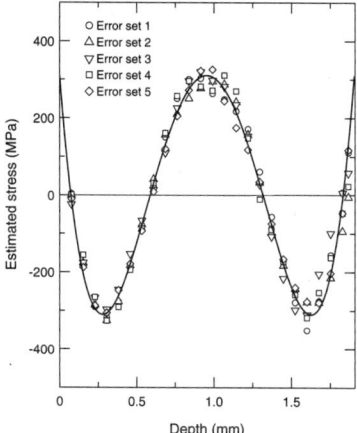

Fig. 6.21. The same as Fig. 6.20 for estimation using quadratic piecewise functions.

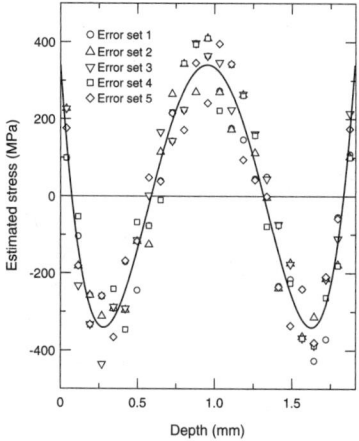

Fig. 6.22. The same as Fig. 6.20 for estimation using the layer removal method.

the "exact strain" ϵ_e to produce the "measured strain" ϵ. Five sets of different random errors tabulated in Table 6.6 are used in the evaluation. The fourth order stress field is estimated again by using the same configurations as those used earlier. The stress distributions estimated using linear piecewise functions and quadratic piecewise functions are shown in Figs. 6.20 and 6.21 respectively.

A similar numerical simulation is also carried out for the layer removal method to estimate the same residual stress distribution. Since the computational model for layer removal only applies to a finite body, we arbitrarily choose a beam with a thickness ten times of the depth of the surface stress measurement. The change of strain due to layer removal from the front surface is computed for a strain gage on the back surface of the beam. The random errors given in Table 6.6 are then superimposed on the computed strains. The estimated stress distributions are shown in Fig. 6.22.

A comparison of Figs. 6.20 and 6.21 with Fig. 6.22 shows that the compliance method using both linear and quadratic overlapping piecewise functions leads to an estimation of the original stress more stable than that using the layer removal method. It is expected that the influence of error on the layer removal method would become even larger for a thicker specimen.

6.6 Configurations Analyzed by the Compliance Method

The crack compliance method was first used to measure axisymmetric stress distributions in thin-walled cylinders. Excellent validation was obtained for a single-pass weld [11] while multi-pass weld results [16] showed good general agreement with values in the literature. The method was then extended to measure residual hoop stress in rings or cylinders by introducing an axial slit of increasing depth [15]. An example of experimental measurement of the hoop stress in a quenched cylinder compared to numerical computation, and surface stress measured by X-rays is shown in Fig. 6.23. Additional development on measurement of the axisymmetric plane strain residual stresses is present in the Chapter 8.

The first application of the crack compliance method to other than axisymmetric geometries dealt with measurement of near surface stresses in a semi-infinite solid [34]. A geometry similar to the front part of Fig. 4.1 was considered with the back surface at infinity. This solution appears to be considerably more sensitive than hole drilling methods for surface stress measurement and permits measurement of stresses below the surface. As was pointed out earlier, the basic theory applies to cracks but in practice a slit of finite width has to be introduced. Solutions [24] for the strains on the front surface at location $y = s$ shown in Fig. 4.1 have been obtained for finite width slits. These solutions greatly extend the applicability of the method for measurement of near-surface stresses. For near-surface stresses it is important to locate the strain gage near the cut. In this case, because of the rapid variation of strain with distance from the cut, we compute the average strain over the gage length of the strain gage. Figure 6.24 shows predictions, using solutions given in Chapter 3 and overlapping-piecewise functions for stresses induced by laser cladding of stellite on steel [31]. General agreement with X-ray measurement is very satisfactory. It is of interest to note that the slit compliance technique is capable of measuring a stress distribution that varies rapidly below the interface between cladding and base metal.

The measurement of stresses on a plane which starts at the toe of a fillet weld or bracket welded to a plate is discussed in [22, 25]. This measurement, which is impractical with layer removal methods (see [126]) and difficult to obtain with X-ray techniques [94], is readily obtained with the slit compliance

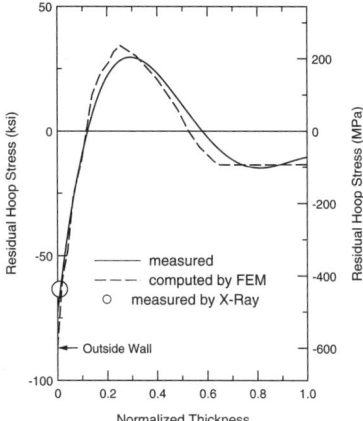

Fig. 6.23. The residual hoop stresses in a water-quenched cylinder measured by the crack compliance method agrees well with the X-ray measurement near the surface.

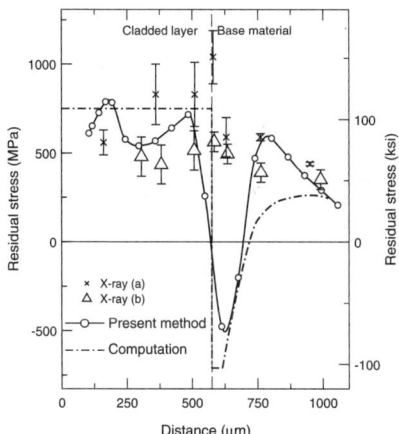

Fig. 6.24. Residual stresses obtained by X-ray method from two independent laboratories (data points), the crack compliance method (solid data line) and numerical simulation (dotted line).

method. Figure 6.25 shows the residual stress distribution measured on the plane near the toe of a bracket-plate assembly.

Two cutting techniques which greatly extend the range of applicability of our approach are electric discharge wire machining (wire EDM) [38] and electric discharge machining (EDM). Cuts can be made with wires as thin as 25 μm which leads to cuts which, even after the kerf width is included, are far thinner than can be achieved with conventional machining. Residual stress due to shot-peening is typically formed very close to the surface. A cut produced

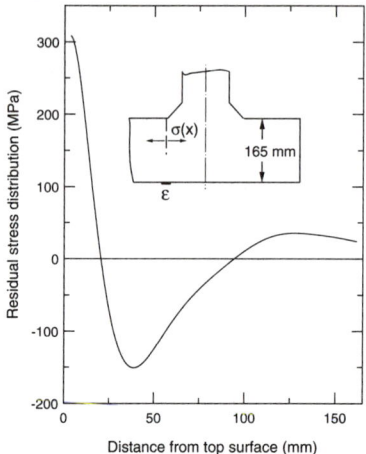

Fig. 6.25. Residual stress on a plane at the toe of the weld between a plate and a bracket.

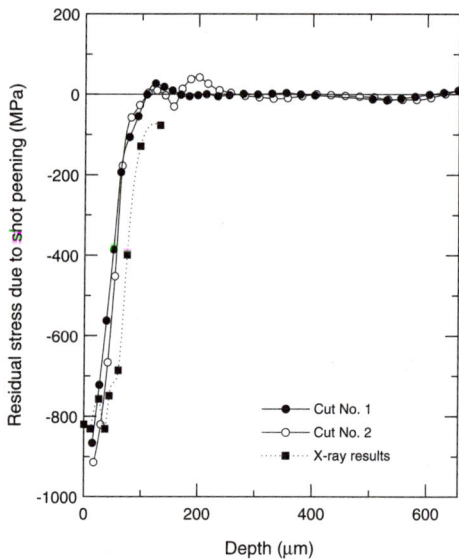

Fig. 6.26. Residual stresses due to shot-peening measured by the slitting method and X-ray technique.

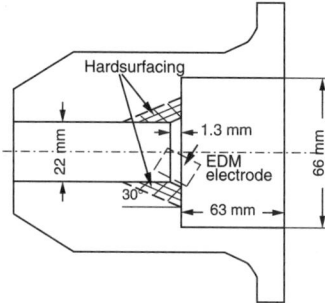

Fig. 6.27. Residual hoop stress in the hardsurfacing is measured directly without any prior machining.

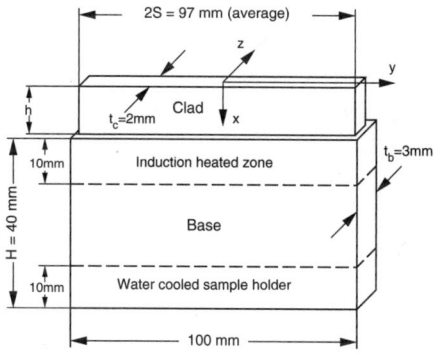

Fig. 6.28. Residual stress distribution in a specimen with non-uniform thickness can be obtained by using a piecewise function through-the-thickness.

by such a thin wire allowed the measurement of residual stress very close to the surface, as shown in Fig. 6.26. Using a specially made electrode, cuts of almost uniform depth can be introduced on a curved surface. Measurements can now be made at locations inside a valve body as shown in Fig. 6.27 where other techniques would not be feasible to apply without cutting the specimen apart.

The use of finite element computation allows compliances to be obtained for parts that are made of different materials and/or are nonuniform in width. Figure 6.28 shows such an example, in which a piecewise function is used to estimate the residual stress that is discontinuous due to the nonuniform width and the change in material properties through the thickness [54]. Additional experimental results have been reported in [2, 33, 45, 48, 73, 88, 91].

6.7 Conclusion

We have presented two different approaches for estimation of residual stresses based on a continuous or a piecewise approximation. For the former one we showed that the use of a Legendre or Chebyshev series leads to a greatly improved condition of the compliance matrix over that of a power series. Unlike the approach that uses a continuous function for which the accuracy of estimation is improved by taking an increasing order of approximation until a convergent result is reached, the approach that uses a piecewise function improves its accuracy of estimation by refining the size of the intervals while the order of approximation, usually $2nd$ order or less and given by power functions in each interval remains constant. This is especially useful for measurement of near surface residual stresses due to surface treatment that leads to a steep stress gradient. For residual stresses that are affected by the discontinuity of the material properties, the use of the piecewise approximation is mandatory and the division of the intervals must match as closely as possible the interface along the continuity.

7

Measurement of Through-Thickness Residual Stress

7.1 Introduction

Measurement of residual stresses through thickness using the layer-removal or sectioning method is traditionally a very time-consuming process, which often makes obtaining a large number of data impractical. For localized residual stresses such as due to welding, the measurement may become more tedious and prone to error in both measurement and analysis. Fortunately, the slitting method, when combined with wire EDM, makes the experimental procedure straightforward, requiring only a single strain gage installed on the surface opposite the cut. A large amount of data can be recorded as the depth of cut is being extended incrementally with high precision wire EDM. The basic assumptions of the method are that the stress does not vary in the transverse direction, and the deformation due to cutting is linear elastic. In this chapter we will present the procedures to be used in measurement of the through-thickness residual stresses.

7.2 A Case Study: Through-Thickness Residual Stress in A Beam Due to Bending

A classic example of residual stresses in textbooks on strength of materials is a beam subjected to bending beyond its elastic limit. Such a test is illustrated in Fig. 7.1, in which a four-point bending fixture [79] produces a pure bending in the middle section of the beam, and the strains recorded on the top and bottom faces as a function of loading allow a stress-strain curve as well as a residual stress distribution after unloading to be obtained in a single test. The availability of the "exact" stress makes it an ideal test for studying the procedures for estimating the through-thickness residual stress [101].

The measurement starts with installing a strain gage on the mid-section of a beam with residual stress generated by a four-point bending test. A wire

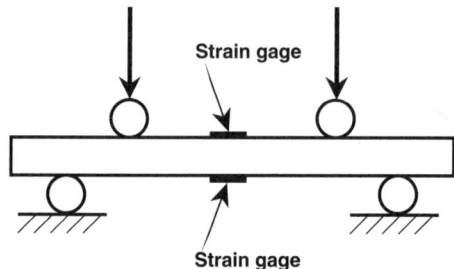

Fig. 7.1. Schematics of a beam bent by a four-point bending fixture while the deformation is recorded by strain gages.

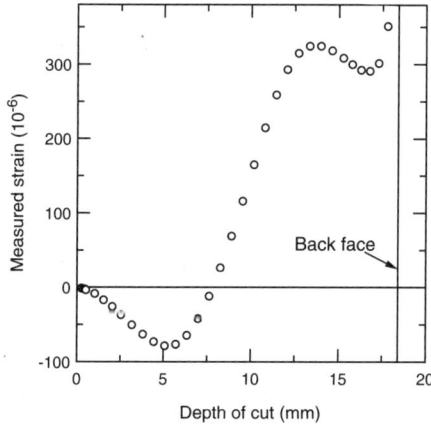

Fig. 7.2. Strain variation measured by the compliance method for a beam bent beyond its elastic limit.

EDM with wire diameter $= 0.0508$ mm is used to make a cut on the front face directly opposite the strain gage. Before cutting, the wire is moved to the back face and positioned to align with the center of the strain gage. The high precision of the wire EDM ensures an excellent alignment of the cutting path with the center of the strain gage without any additional adjustment. The strain measured during cutting is plotted against depth of cut in Fig. 7.2. Table 7.1 tabulates the geometry and material properties of the specimen and the configuration of the measurement. A subtle but important issue is to determine E' which is only known for two special cases: $E' = E$ for plane stress and $E' = E/(1-\nu^2)$ for plane strain. Although most parts are not wide enough for the deformation to be plane strain when a cut is just introduced, the deformation will eventually become one in plane strain as the ratio of width to thickness of the remaining ligament increases. Fortunately, the difference between the two extreme cases is not very significant, about 9% for $\nu = 0.3$. As an approximation, we choose the mean between E and $E/(1 - \nu^2)$, i.e., $E' = E/(1 - \nu^2/2)$, which will lead to at most an error about 4.7% for very shallow and deep cuts.

Table 7.1. Configuration of the Measurement

Beam Geometry			Stainless Steel 304L		Measurement Configuration		
Width (mm)	Thickness (mm)	Length mm	E (MPa)	ν	Cut width (mm)	Gage location	Gage length (mm)
18.415	18.415	152	196	0.3	0.157	Back face	0.8128

The width of the cut is seen to be very small compared to the thickness and the cut can be approximated by an extending crack. We will first use the LEFM approach to compute the crack compliances. Since the compliance functions can be conveniently computed over a large number of depths of equal increment, which often differs from the depth/increment of the cut used in a measurement, an interpolation procedure is usually required to find the values of compliance functions at the depths of measurement or the values of the strain at the depths of the compliance computation. The first approach limits the number of data used in stress estimation to that of strain measurements while the second approach may have to give up the first and last strain data located at the ends of interpolation. To overcome these limitations, we combine all the data points found in both approaches in the stress estimation that follows. An enlarged data set will improve the condition of the resulting compliance matrix but not enhance the quality of the data nor reduce the influence of measurement error.

After the strain data and compliance functions are determined at the same depths of cut, a least squares fit (LSF) is carried out for approximations of increasing order from $n = 2$ to $n = 19$ using a complete or a truncated Legendre series. Figures 7.3 and 7.4 show the results. It is seen that the estimation in the region $0.2 < a/t < 0.9$ converges consistently for both series while the estimated stress in the regions of $a/t \leq 0.2$ and $a/t \geq 0.9$ oscillates as the order of approximation increases. This observation suggests that a convergence test can be performed over the data in the region of $0.2 < a/t < 0.9$, which we refer to as the main convergence test. Once the main convergence test is satisfied, stability in the regions of $a/t \leq 0.2$ and $a/t \geq 0.9$ can be improved by taking the average. To quantify the main convergence test we define a normalized deviation over two consecutive orders of estimation, i and $i - 1$, as

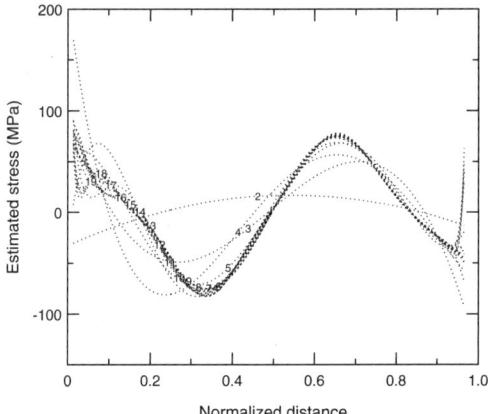

Fig. 7.3. Stress approximated by a complete Legendre series of increasing orders obtained by LSF.

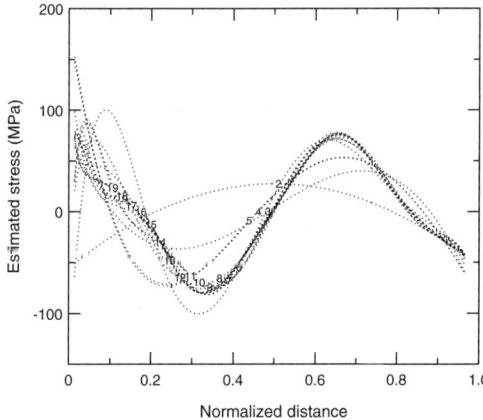

Fig. 7.4. Stress approximated by a truncated Legendre series of increasing orders obtained by LSF.

$$\delta_i = \sqrt{\frac{\sum_{k=1}^{m}[\sigma_i(x_k) - \sigma_{i-1}(x_k)]^2}{\frac{1}{4}\sum_{k=1}^{m}[\sigma_i(x_k) + \sigma_{i-1}(x_k)]^2}} \qquad 0.2 < x_k < 0.9 \qquad (7.1)$$

Figure 7.5 shows the plots of the main deviation δ_i as a function of i. It is seen that convergence is achieved for both cases when δ_i becomes less than about 6%. As a simple rule, thus, the convergence is assumed to start when δ_i becomes and remains 6% or less for all higher order estimations that follow.

Once the estimation is found to converge, say at order j, the stress estimated at order $n \geq j + 1$ is averaged according to

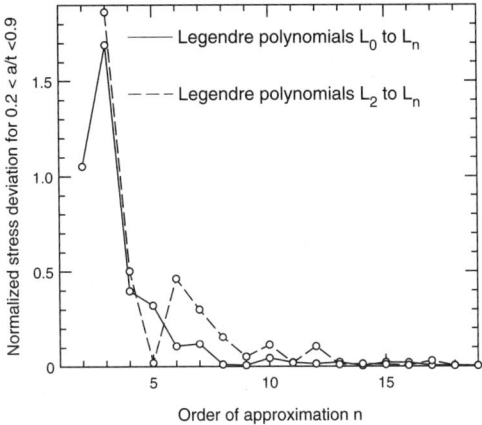

Fig. 7.5. δ_i for a complete and a truncated Legendre series.

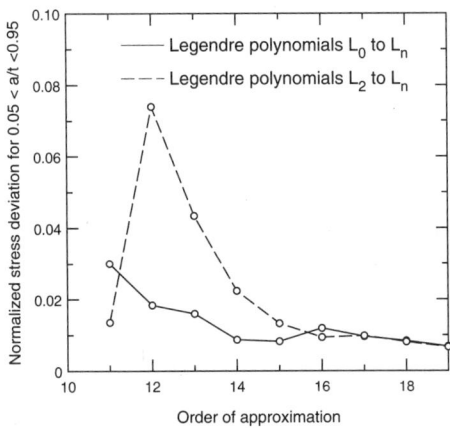

Fig. 7.6. Δ_i for a complete and a truncated Legendre series.

$$\sigma_n^a(x) = \frac{1}{n - j + 1} \sum_{k=j}^{n} \sigma_k(x) \tag{7.2}$$

This process applies increasing damping to the higher order terms without modifying the terms less than or equal to j, and an overall convergence is now possible. We use the same computation to obtain the overall deviation Δ_{n+1} on the averaged estimations over a range of $0.05 < a/t < 0.95$,

$$\Delta_{n+1} = \sqrt{\frac{\sum_{k=1}^{m} [\sigma_{n+1}^a(x_k) - \sigma_n^a(x_k)]^2}{\frac{1}{4} \sum_{k=1}^{m} [\sigma_{n+1}^a(x_k) + \sigma_n^a(x_k)]^2}} \qquad 0.05 < x_k < 0.95$$

The overall convergence is considered having been reached when two consecutive averaged estimates become less than 4%. The latter of the two is then

taken as the final estimated residual stress. We refer to the convergence test
for the averaged stresses as the overall convergency test. Figure 7.6 shows
the plots of Δ_n as a function of n. It is seen that the overall convergence is
achieved quickly for both cases. The final estimated residual stress distributions are shown in Fig. 7.7.

When a weighted LSF is used, the above procedures can be applied equally.
Because the order of the main convergence is found to occur at $n > 8$, from
Fig. 6.10 we choose the exponent $\zeta = 1.2$ in the weight function, and the
estimated final stresses are also shown in Fig. 7.7. Comparison of the stress
distributions in Fig. 7.7 shows that the weighted LSF indeed improves the
result in the region of $a/t < 0.2$.

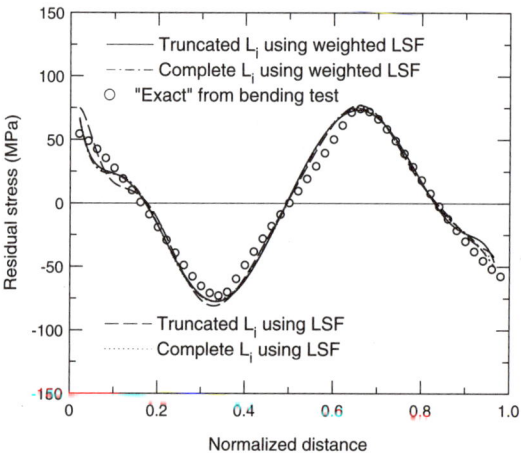

Fig. 7.7. Comparison of the residual stresses estimated using the LEFM approach
with the "exact" result computed from the bending test.

We now turn to the compliance functions computed by FEM with 216
8-node elements along the plane of cut. Using the same procedures, the stresses
are estimated and shown in Fig. 7.8. Again the weighted LSF is found to improve the estimation. It is interesting to note, in contrast to the difference
shown in Fig. 6.9, that the result is very similar to that obtained using compliance functions computed by LEFM for complete or truncated Legendre
series with or without weighting in LSF. This shows that when the residual
stress variation is dominated by higher order terms $(n > 4)$ the difference in
stresses estimated using the two different approaches is very small.

7.3 Dominant Variation in Stress Estimation

When a polynomial series is used to estimate residual stresses, a key issue is to
decide what order of an approximation is sufficient to represent the unknown

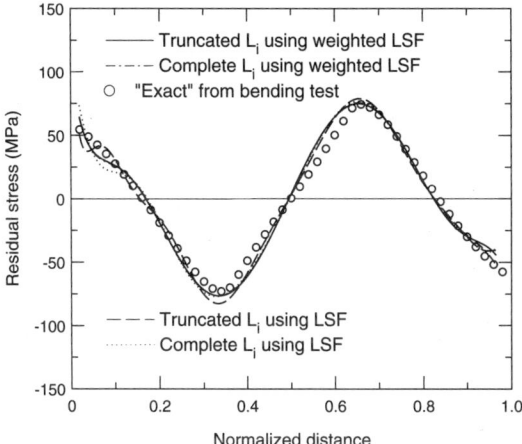

Fig. 7.8. Comparison of the residual stresses estimated using the FEM approach with the "exact" result computed from the bending test.

residual stress. Another issue of equal importance is how to achieve stability for high order approximations given the unknown nature of the error involved in measurement and computation. During the estimation, using increasingly higher orders, we made two observations: a convergence can be achieved consistently in the region of $0.2 < a/t < 0.9$, and the estimated stress in the regions of $a/t \leq 0.2$ and $a/t \geq 0.9$ shows considerable oscillation. The first observation suggests that a sufficient approximation can be reached by the convergence of the estimation as the order of approximation increases. The second observation indicates that the stability of estimation in the regions of $a/t \leq 0.2$, and $a/t \geq 0.9$ can be improved by taking the average. Note that taking the average should be made only after the main convergence test has been satisfied. Otherwise, prematurely applying the process of averaging will make a sufficient approximation unattainable.

The procedure of the main convergence test assumes that the stress approximation has captured the dominant variation of the stress after the criterion for δ_i has been met by all subsequent orders of estimation. To quantify the meaning of dominant variations, let us consider a residual stress constructed by two groups of Legendre polynomials,

$$\sigma(x, \alpha) = [10L_3(x) + 15L_4(x)] + \alpha[10L_8(x) + 15L_9(x)] \qquad (7.3)$$

Through a numerical computation we can simulate the strain variations measured due to the release of the stresses with decreasing values of α and identify the onset of the first group becoming dominant on the final stress estimation. Table 7.2 tabulates the estimated coefficients for different strain variations computed by LEFM for $\alpha = 0.15, 0.1, 0.07$ and 0.05. It is seen that for α approaching 0.07 the main convergence test becomes satisfied after the first group is included in the estimation, and the computation of the stress averaging continues for next three orders. That is, the first two terms become dominant when $\sigma(1, \alpha = 0)/\sigma_{est}(1, \alpha)$ is about 0.95 from Table 7.2. Thus, we refer to the ratio of the estimated peak stress when the main convergence is reached to that of the original stress as the dominant variation ratio, R_d, which increases with the decrease of the tolerance specified for the main convergence test and *vice versa*. The ratio may serve as an estimate of the minimum accuracy achievable for the measured stress variation for a given main convergence test. The phrase *minimum achievable* is used because it corresponds to the situation when the error due to omission of higher order terms that satisfy the convergence test happens to be maximum. For the present example, it implies that the achievable accuracy is at least 95%. Although an estimate of the minimum achievable accuracy is useful for establishing the lower bound of the measurement accuracy, the influence of other independent variables, such as the error in specified value of E' and those to be discussed in the next section, should also be included in evaluation of the final accuracy.

Table 7.2. Coefficients in Eq. (7.3) Estimated Using Weighted LSF for Different Values of α

Coef. order	$\alpha=0.15$		$\alpha=0.1$		$\alpha=0.07$		$\alpha=0.05$	
	$\sigma(x,\alpha)$	$\sigma_{est}(x,\alpha)$	$\sigma(x,\alpha)$	$\sigma_{est}(x,\alpha)$	$\sigma(x,\alpha)$	$\sigma_{est}(x,\alpha)$	$\sigma(x,\alpha)$	$\sigma_{est}(x,\alpha)$
2	0.0	0.0	0.0	0.0	0.0	0.00003	0.0	0.00002
3	10.0	10.0	10.0	10.0	10.0	9.99833	10.0	9.9988
4	15.0	15.0	15.0	15.0	15.0	15.04224	15.0	15.0302
5	0.0	0.0	0.0	0.0	0.0	-0.01814	0.0	-0.01296
6	0.0	0.0	0.0	0.0	0.0	0.02887	0.0	0.02062
7	0.0	0.0	0.0	0.0	0.0	0.01996	0.0	0.01426
8	1.5	1.5	1.0	1.0	0.7	0.53689	0.5	0.38349
9	2.25	2.25	1.5	1.5	1.05		0.75	
$\sigma(1,\alpha)$	28.75	28.75	27.5	27.5	26.75	25.6082	26.25	25.4344
σ_{est}/σ		100%		100%		95.7%		96.9%

Different residual stress distributions will lead to different values of dominant variation ratio. In practice, an approximate dominant variation ratio can be estimated after a convergent stress estimation has been obtained. Denoting the stress estimated when the main convergence is reached at order j as $\sigma_{mj}(x)$, the stress field to be used for evaluation of the dominant variation ratio may be constructed as

$$\sigma(x, \alpha) = \sigma_{mj}(x) + \alpha[\sigma_{mj}(x_p)L_{j+3}(x)] \tag{7.4}$$

where $0.2 < x_p < 0.9$ is the distance at which stress $\sigma_{mj}(x)$ reaches its peak value. Equation (7.4) is expected to be valid as long as the magnitude of the higher order $(> j + 3)$ terms in the original residual stress distribution diminishes as the order increases. Carry out the same computation as that used for Table 7.2; the dominant variation ratio can be obtained.

For a concrete example we turn back to the residual stress due to bending. For estimation using truncated Legendre series with weighted LSF shown in Fig. 7.7, the order at which the main convergence occurs is $j = 9$. Table 7.3 lists the results computed for decreasing values of α. It is seen that $\sigma_{mj}(x)$ becomes dominant when $\alpha \leq \alpha_d = 0.321$. The value of the dominant variation ratio, R_d, is found to be about 0.93. Note that α_d obtained for the residual stress due to bending is considerably larger than that found for the stress in Eq. (7.3). The difference is not unexpected because Eq. (7.4) uses the peak stress $\sigma_{mj}(x_p)$ as the amplitude for the second term while Eq. (7.3) uses the surface stress $(x = 1)$.

Table 7.3. Results of Evaluation Using Eq. (7.4) for Different Values of α

values of α	0.559	0.349	0.335	0.321	0.279	0.140
Main convergence order j	12	12	12	9	9	9
$\sigma_{est}(x_p, \alpha)/\sigma(x_p, \alpha)$	1.0	1.0	1.0	0.930	0.939	0.969

We now can estimate the error bound on the peak variation, Δ_p using the dominant variation ratio, i.e.,

$$\eta_p = \pm(1 - R_p)\sigma_{mj}(x_p) \tag{7.5}$$

Figure 7.9 shows the error bound estimated for the measured stress due to bending. The prediction is seen to be satisfactory for the peak variations. It, however, underestimates the error near the front and back faces. A more detailed error analysis in these regions will be presented in the next section.

7.4 Error in Through-Thickness Measurement

Having determined the residual stress, an analysis of the error in estimated stress is of great practical importance. However, the wide range of variables involved in a residual stress measurement usually make it impossible to have a single approach that can cover all the possibilities. For example, the approach using the dominant variation ratio appears applicable to the error bound for the peak variations in the middle region but not to the regions near the surface.

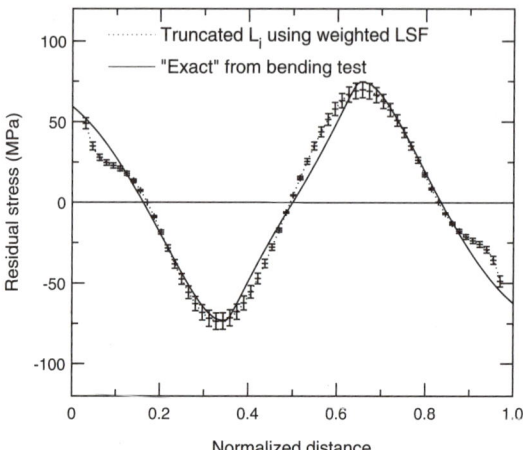

Fig. 7.9. Error bound estimated from dominant variation ratio R_p.

Also, there always exists an error that happens to contain a component that coincides with a strain variation produced by a term used in the residual stress approximation. Such an error will be difficult to detect by an error analysis using a single set of data. We may refer to this kind of error as the shape error because it mainly changes the shape of the estimated stress distribution. Aside from this limitation, we may assume that an error-free estimation is reached when the estimated stress remains unchanged as the order of approximation increases. This, of course, is unlikely to occur in most practical applications but it does suggest that the error is related to the changes in estimated stresses as the order of approximation increases. In other words, after the main convergence is reached, the oscillation observed in the regions near the surface are mostly due to the influence of the error randomly distributed in the measured strain. Denoting the final averaged stress obtained using Eq. (7.2) as σ_n and the stress estimated at order $k \geq j$ as σ_k, the error bound based on the oscillation of the estimated stresses, η_e, may be written as

$$\eta_e(x) = \pm \max_{k=j}^{n+2}[\|\sigma_k(x) - \sigma_n(x)\|] \tag{7.6}$$

in which the maximum variation among stresses of increasing order is chosen as the error bound. In Eq. (7.6) we carry out the selection from $k = j$ until $k = n + 2$, two orders above the order of the averaged final stress. Using a larger number of the stress approximations could provide a more complete coverage of the stress bound over the entire region but it may also lead to an overestimation of the error bound if the estimation starts to diverge as the order of approximation increases.

Since $\eta_e(x)$ in Eq. (7.6) is independent of the error bound of the peak variation η_p in Eq. (7.5) the combined error bound $\eta(x)$ may be given as

$$\eta(x) = \pm\sqrt{\eta_e^2(x) + \eta_p^2} \qquad (7.7)$$

Figure 7.10 shows a prediction of the combined error bound for the measured residual stress due to bending. Although the improvement in the region near the back face is moderate, the improvement in the region near the front face is considerable compared to that shown in Fig. 7.9.

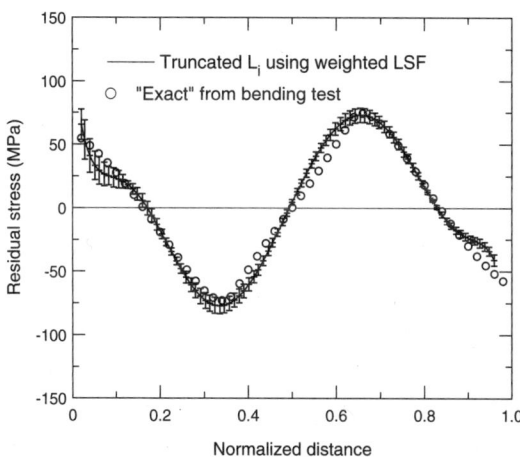

Fig. 7.10. Combined error bound $\eta(x)$ estimated from Eq. (7.7).

Besides the randomly distributed error in the measured strain and the error due to omitting higher order terms after the main convergence is reached, there is another kind of error η_l that resembles a strain variation produced by an unbalanced load. Unlike the random error, η_l always increases monotonically with depth of measurement and becomes very significant near the back surface. Its likely sources are the error in computed compliance functions, influence of the specimen's weight, and effect of cutting produced by wire EDM.

Consider a stress given by a modified Gaussian distribution, which was originally used by Prime and Hill [100],

$$\sigma(x) = 2.28735 \left\{ \alpha \left[0.1811423 - 0.4286826 L_1(x) \right] - e^{-\left(\frac{x-0.05}{0.15}\right)^2} \right\} \qquad (7.8)$$

where the value of α determines the magnitude of the resulting unbalanced load and for $\alpha = 1$ $\sigma(x)$ becomes essentially a residual stress, which can not be expressed exactly by a finite series expansion. Thus, the estimated stress will always include the model error caused by a finite order approximation.

First, we use LEFM to generate strain variations to be used for stress estimation, as shown in Fig. 7.11, and compliance functions up to the 19th order. It is seen that even a small amount ($\Delta\alpha = 0.025\%$) of unbalanced loading leads to a considerable change in strain for deep cuts. Next, using a truncated Legendre series to approximate the stress, a weighted LSF is carried out for the four strain variations. The results are plotted in Fig. 7.12. For $\alpha = 1$ the estimation is almost a perfect fit. This indicates that a convergent estimation is essentially free of the influence of model error. For $\alpha \neq 1$ the error in estimated stress is seen to increase as the amount of the unbalanced load increases and is much more noticeable near the front and back faces. For

Table 7.4. Results of Estimation Using Different Approximations for Eq. (7.8)

values of α	1.0	1.00025	1.0005	1.001	Legendre series
Main convergence order[†] j	8	11	16	16	$L_i(x)$ $i \geq 2$
Main convergence order[‡] j	8	8	8	8	$L_i(x)$ $i \geq 0$

† Weighted LSF with $\zeta = 1$ ‡ LSF or weighted LSF

$\alpha = 1.00025$ the combined error is estimated and plotted in Fig. 7.13. The overall agreement is seen to be very satisfactory. For $\alpha \geq 1.0005$, however, the order of main convergence jumps to $j = 16$, which is very close to the highest order ($n = 19$) of compliance functions used in the evaluation to make a complete error analysis.

Alternatively, using a complete Legendre series and weighted LSF, estimation is carried out for the same three strain variations with $\alpha \neq 1$. In this case the influence of η_l is absorbed by the uniform and linear terms of the polynomial series, and the estimated residual stress is found basically independent of α. Thus, the first two terms of the complete Legendre series provide a reasonable estimate of η_l. If the measured strain is subjected to only a very small amount of η_l, the use of truncated Legendre series will lead to a better estimation especially when a higher order approximation is required. Figure 7.14 shows a comparison of the stresses estimated near the front surface where the error is found to be largest. The result obtained using a truncated Legendre series is seen to be excellent for $\alpha = 1$ but sensitive to a small change in α while the use of a complete Legendre series makes the result virtually independent of α.

The evaluation of η_e in Eq. (7.6) is based on a necessary condition for an error-free stress estimation. It may not cover errors under other necessary conditions, for example, the change in stress estimated by using an alternative set of data. When errors η_k are obtained from m independent sources, the combined error is usually expressed as

$$\eta(x) = \pm \sqrt{\sum_{k=1}^{m} \eta_k^2(x)} \tag{7.9}$$

which is a generalized form of Eq. (7.7).

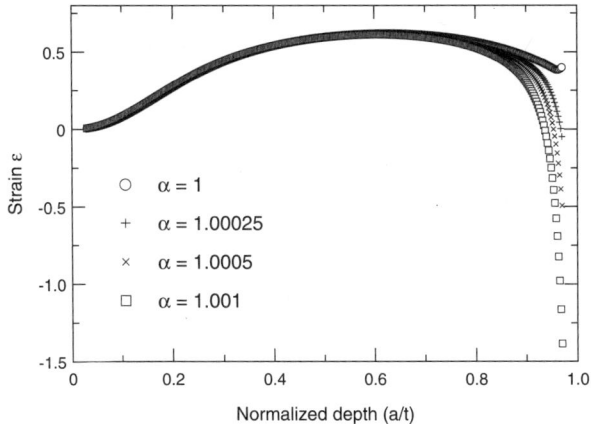

Fig. 7.11. Strain generated by LEFM from a modified Gaussian function with a varying amount of unbalanced loading.

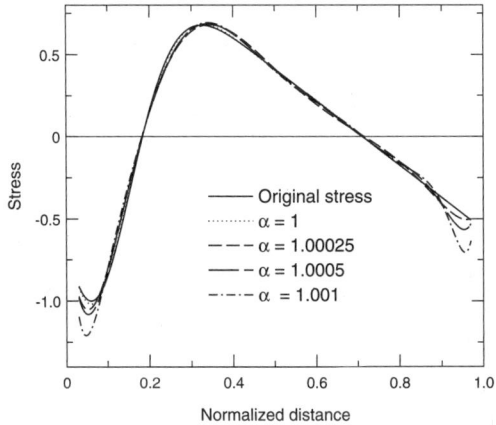

Fig. 7.12. Stress estimated from strain variations given in Fig. 7.11.

7.5 Conclusion

In the experimental study of a through-thickness measurement we found that convergence in the middle $60-70\%$ portion of the thickness can be reached for a continuous residual stress as the order of approximation increases. A general approach is then developed to determine the order of approximation which will result in a convergent and stable estimation. The procedures presented will be applicable to the measurement of residual stresses that are continuous through the thickness except about $2-5\%$ near the front and back surfaces.

The simple error analysis developed here appears to give a reasonable prediction of the error bound of the estimated stress. Through numerical simulations we also demonstrate that the effect of error that corresponds a

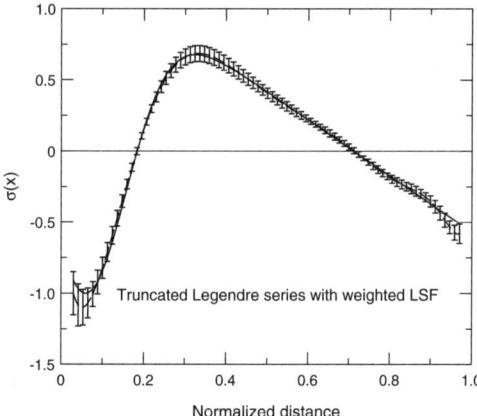

Fig. 7.13. Stress and error bound estimated from strain variation with $\alpha = 1.00025$ in Eq. (7.8).

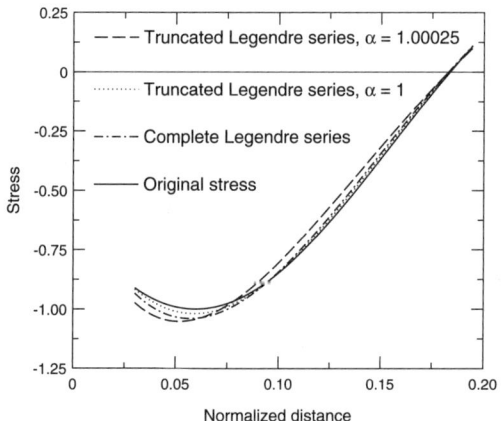

Fig. 7.14. Comparison of the stresses estimated near the front surface where error reaches maximum.

strain variation produced by an unbalanced load can be greatly reduced by using a complete series expansion.

8

Measurement of Axisymmetric Residual Stresses

8.1 Measurement Using Two Axial Cuts

8.1.1 Introduction

The traditional approach for measurement of axisymmetric residual stresses in a hollow cylinder in plane strain involves removing annular layers from the inside surface while measuring axial and hoop strains at the outside surface. Although normally attributed to Sachs (1927)[113], the equations required for this approach were first developed by Mesnager (1919)[82]. He also pointed out that deformation could be measured at the inside surface while annular layers were removed from the outside. Work prior to Mesnager by Heyn and Bauer (1910)[64] only considered axial residual stresses. For a solid rod, the Mesnager-Sachs procedure requires that a central core be removed before layer removal is carried out. Strain measurements on the outside diameter allow the effect of core removal on the stresses in the cylinder to be computed. However, only average values of the stresses in the core region can be obtained.

The layer removal procedure is extremely time consuming experimentally, but the subsequent reduction of strain measurements to determine stresses is quite straightforward. Perhaps for this reason it continues to be used to the present day.

Besides the boring-out method, an ingenious and more precise method for residual stress measurement has been described by Ueda and his colleagues (1986)[128]. This involves separating a circular disk and a rectangular plate cut in the axial direction from the cylinder. Strain gages are attached to the disk and plate which are then sectioned to obtain the data required for the estimation of residual stresses. While less time consuming than the layer removal method, this procedure requires a fairly large number of strain gages and a considerable amount of machining.

In Sections 4.3 and 4.5 we described the slitting method for measurement of hoop stresses in disks, thick-walled cylinders or rings. This method requires only the measurement of the hoop strain at one location while a thin cut of

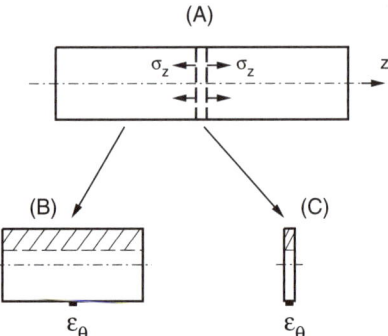

Fig. 8.1. An axisymmetric residual stress in plane strain is measured by an axial cut made on a cylinder/rod and another one made on a ring/disk cut out from the original cylinder/rod.

progressively increasing depth is made in the axial-radial plane through the thickness. However, the procedure described in Section 4.4 for measurement of axial stress by introducing a circumferential cut applies only to a thin-walled cylinder, and the cutting is more difficult to make than that of a radial cut.

With this background we will present two methods which will further simplify the measurement of axisymmetric residual stresses. First, we will show that measurements of hoop stresses in a thin ring or a disk and a cylinder with a length more than twice its diameter, both of which are cut from a long cylinder, may be combined to predict also the radial and axial stresses in the long cylinder. The theoretical basis for the approach is quite general and is not restricted to the slitting method. The procedure will then be used to estimate the radial and axial residual stress distributions in a water quenched cylinder using hoop stresses measured for a disk and a cylinder.

Next, we will show how to use the deformation measured when a single slice is removed from the middle section of a cylinder and the remaining residual hoop stress in the slice to obtain both the hoop and axial residual stresses in the original cylinder. In this case the required length of the original cylinder is substantially reduced.

8.1.2 Analysis of Axisymmetric Residual Stresses in Plane Strain

Consider a cylinder of uniform geometry in the z-direction, as shown in Fig. 8.1(A), which is free of any external loading. The elastic properties of the body, namely Poisson's ratio ν and elastic modulus E, are assumed to be homogeneous and isotropic. If the length of the cylinder is more than twice its diameter, the axisymmetric residual stresses σ_r, σ_θ and σ_z in the middle portion of the cylinder are in plane strain. We now cut out a slice (a disk or a ring) from the middle portion of the cylinder. Since the axial stress σ_z on the plane of cut is completely released, the residual stresses remaining in the slice are in plane stress. Denoting the change of stresses due to releasing σ_z with a prime and the stresses in the slice with a superscript s, the stresses in the original cylinder can be expressed as

$$\sigma_r = \sigma_r^s + \sigma_r', \quad \sigma_\theta = \sigma_\theta^s + \sigma_\theta' \quad and \quad \sigma_z = \sigma_z' \tag{8.1}$$

The hoop stresses σ_θ and σ_θ^s can be obtained with the crack compliance method, presented in Section 4.3 for a disk and Section 4.5 for a ring, by making an axial cut of progressively increasing depth on the cylinder or slice as shown in Fig. 8.1(B) and Fig. 8.1(C). The radial stress σ_r can be obtained from the hoop stress by using the equilibrium equation [124]

$$\frac{d\sigma_r}{dr} + \frac{\sigma_r - \sigma_\theta}{r} = 0 \tag{8.2}$$

from which the radial stress can be expressed as

$$\sigma_r = \frac{1}{r}\left(\int_a^r \sigma_\theta dr - \int_a^b \sigma_\theta dr\right) \tag{8.3}$$

in which a and b are the inner and outer radii of the cylinder and r is some intermediate radius. Noticing that σ_r is finite at $r = 0$ for a solid cylinder and is zero at $r = a$ for a hollow cylinder, the second term of the above equation must vanish, i.e.

$$\int_a^b \sigma_\theta dr = 0 \tag{8.4}$$

which leads to

$$\sigma_r = \frac{1}{r}\int_a^r \sigma_\theta dr \tag{8.5}$$

The radial stress σ_r^s may also be obtained from Eq. (8.5) by replacing σ_θ by σ_θ^s.

We now turn to the stresses due to releasing the axial stress, which satisfy the linear elastic stress-strain relations without the necessity of including the incompatible initial strains which lead to residual stresses, i.e.,

$$\epsilon_r' = \frac{\sigma_r' - \nu(\sigma_\theta' + \sigma_z)}{E} \quad \epsilon_\theta' = \frac{\sigma_\theta' - \nu(\sigma_z + \sigma_r')}{E} \tag{8.6}$$

Using the relations between the radial displacement and strains

$$\epsilon_r' = \frac{du}{dr} \quad and \quad \epsilon_\theta' = \frac{u}{r} \tag{8.7}$$

we find

$$\epsilon'_r = \frac{d(r\epsilon'_\theta)}{dr} \tag{8.8}$$

Substituting Eq. (8.6) into Eq. (8.8) leads to

$$(1 + \nu)(\sigma'_\theta - \sigma'_r) + r\frac{d}{dr}(\sigma'_\theta - \nu\sigma'_r) = \nu r\frac{d\sigma_z}{dr} \tag{8.9}$$

which when combined with Eq. (8.2) yields

$$\frac{d}{dr}(\sigma'_r + \sigma'_\theta) = \nu\frac{d\sigma_z}{dr} \tag{8.10}$$

or after integration

$$\sigma_z = \frac{1}{\nu}(\sigma'_r + \sigma'_\theta) + C \tag{8.11}$$

in which C is a constant to be determined. Combining Eq. (8.1) with Eq. (8.11) gives

$$\sigma_z = \frac{[(\sigma_r + \sigma_\theta) - (\sigma^s_r + \sigma^s_\theta)]}{\nu} + C \tag{8.12}$$

We now use the condition of no resultant force in the axial direction to show that the constant C is identically zero. Substituting Eq. (8.5) into Eq. (8.12) and integrating the axial stress over the area of the cross section yields

$$0 = \int_a^b \sigma_z r dr = \frac{1}{\nu}\int_a^b [(\sigma_r + \sigma_\theta) - (\sigma^s_r + \sigma^s_\theta)]r dr + C\frac{b^2 - a^2}{2}$$
$$= \frac{1}{\nu}\{\int_a^b [\int_a^r (\sigma_\theta - \sigma^s_\theta)dr]dr + \int_a^b r(\sigma_\theta - \sigma^s_\theta)dr\} + C\frac{b^2 - a^2}{2} \tag{8.13}$$

Carrying out the second integral by part and making use of Eq. (8.4), we find

$$\int_a^b r(\sigma_\theta - \sigma^s_\theta)dr = b\int_a^b (\sigma_\theta - \sigma^s_\theta)dr - \int_a^b [\int_a^r (\sigma_\theta - \sigma^s_\theta)dr]dr$$
$$= -\int_a^b [\int_a^r (\sigma_\theta - \sigma^s_\theta)dr]dr \tag{8.14}$$

which, when substituted into Eq. (8.13), leads to $C = 0$. Now we combine Eq. (8.5) and Eq. (8.12) to obtain the final expression for the axial stress in terms of hoop stress

$$\sigma_z = \frac{1}{\nu}[(\sigma_\theta - \sigma^s_\theta) + \frac{1}{r}\int_a^r (\sigma_\theta - \sigma^s_\theta)dr] \tag{8.15}$$

The above analysis applies equally to a solid cylinder with $a = 0$, when the computational model of the compliance method given in Section 4.3 is used to measure the hoop stresses in a solid cylinder and a disk. It should be noted that Eq. (8.15) applies when there is no external axial loading as in residual stress problems. If this is not the case a constant term must be added to Eq. (8.15).

8.1.3 Determination of the Axial Residual Stress in a Water-Quenched Cylinder

We now use the procedure developed in the previous section to estimate the radial and axial residual stress distributions in a solid Ni-Cr steel cylinder of 1.31 m in length and 0.105 m in radius using data reported by Ueda, et al (1986)[128]. The preparation of the specimen consists of three steps: oil quenching, heating at $870°C$ for three hours and water quenching. The material properties used for residual stress estimation are elastic modulus $E = 206$ GPa and Poisson's ratio $\nu = 0.3$.

In their study Ueda and his colleagues obtained the hoop stresses in the cylinder and a disk cut out from the cylinder. From the hoop stresses we can estimate the radial stresses in the cylinder and the disk. A 9th order power series is used to approximate the hoop stresses by a least squares fit. Then the quantity in the brackets on the right hand side of Eq. (8.15) is obtained and expressed also in terms of a 9th order power series with coefficients given in Table 8.1. The distribution of the axial stress is shown in Fig. 8.2. The overall agreement with the results given by Ueda et al, shown as data points, is very good.

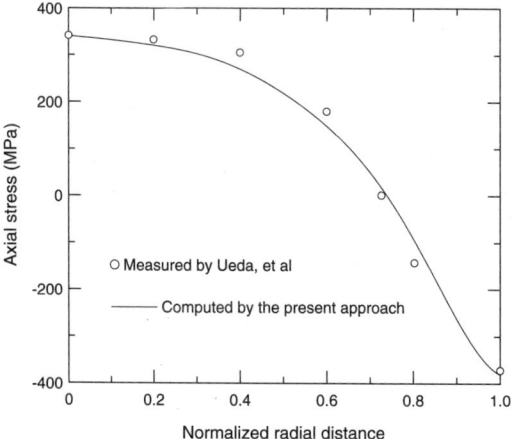

Fig. 8.2. Comparison of residual axial stress distribution computed by the present approach and measured by Ueda, et al[128].

Table 8.1. Coefficients of the power series for $\nu\sigma_z$

$a_0 = 102.07544$	$a_1 = -26.35584$
$a_2 = 175.13632$	$a_3 = -3203.47455$
$a_4 = 22729.36093$	$a_5 = -85770.57992$
$a_6 = 175376.17210$	$a_7 = -197538.08304$
$a_8 = 114779.11670$	$a_9 = -26737.46013$

8.1.4 Discussion

The present analysis is primarily developed for measurement of axisymmetric residual stresses in a long cylinder. It is now possible for the slitting method to estimate an axial residual stress distribution from hoop stresses measured by making one axial cut on a long cylinder and another one on a ring or disk cut out transverse to the cylinder. An implicit assumption of this approach is that the hoop stress in the cylinder is in plane strain. We will examine the condition for plane strain in detail in the next section.

8.2 The Single-Slice Approach for Axisymmetric Stresses

8.2.1 Introduction

In last section we showed that measurements of the hoop stresses in both a long cylindrical body in plane strain and a thin disk cut from it can be used to deduce the axial stress in plane strain. This approach, as illustrated in Fig. 8.1, has the advantage of only requiring straight axial cuts, which in conducting materials may be made conveniently using wire EDM, and may be applied to rods or cylinders. The experimental results of Ueda and his colleagues [128] provided a striking validation of this procedure.

Some practical considerations are that cutting a long, large diameter cylinder with wire EDM is time consuming, although very much less so than the boring-out procedure, and the length required may exceed the capacity of the machine. Also, in cutting brittle metals with high surface tensile stresses in plane strain, the narrow cut produced by wire EDM may propagate spontaneously, which invalidates the test.

With this background we have developed a new method which involves a transverse cut in the central region of the rod or cylinder, as shown in Fig. 8.3. Strain gages are then attached to one of the faces exposed by the cut and then a thin circular slice containing the gages is removed by another transverse cut. From the change in strains due to removal of this slice, the original axial

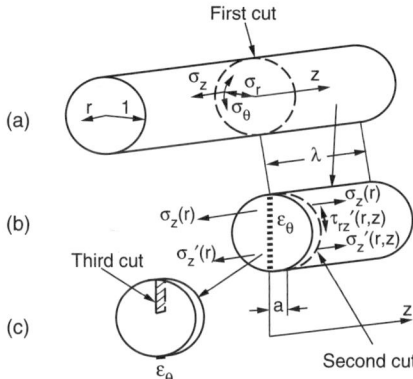

Fig. 8.3. Stress states involved in single-slice method: (a) original rod, (b) after first cut, and (c) a slice is removed.

stress in plane strain may be obtained. By attaching a single strain gage to the surface of the slice and machining a diametral cut of progressively increasing depth, the hoop stress in plane stress may be deduced. Using expressions presented, the original hoop and radial stresses in plane strain may be obtained from the hoop stress in the slice and the axial stress computed from the slicing technique.

If crack propagation should occur during the first cut due to high tensile stresses, another transverse cut may be made close to the fracture surface before installing strain gages. The axial stress in the region of the second cut will be much smaller than that for the first cut.

8.2.2 Estimation of the Axial Residual Stress

For convenience in the derivation that follows, all dimensions are normalized by the outside radius of a solid rod or by the thickness for a hollow cylinder. First, for a rod with a normalized radial distance denoted by r, we express the axial stress in the plane strain region by

$$\sigma_z(r) = \sum_{i=1}^{n} A_z^i S_i(r) \quad with \quad \int_0^1 S_i(r) r \, dr = 0 \qquad (8.16)$$

The functions $S_i(r)$, which we will specify later, must satisfy the equilibrium condition expressed by the second term of Eq. (8.16). The amplitude factors A_z^i are the quantities which have to be obtained from the experimental procedure to determine the original axial residual stress in plane strain $\sigma_z(r)$.

For a cylinder of normalized inner radius b and thickness of unity, Eq. (8.16) would be replaced by

$$\sigma_z(x) = \sum_{i=1}^{n} A_z^i S_i(x) \quad with \quad \int_0^1 S_i(x)(b+x)dx = 0 \qquad (8.17)$$

where x is the normalized radial distance starting from the inside surface.

For simplicity, we discuss the procedure for the solid rod shown in Fig. 8.3-a. A similar derivation applies for a cylinder with $S_i(r)$ replaced by $S_i(x)$ and r replaced by x. The stresses $\sigma_z(r)$, $\sigma_\theta(r)$ and $\sigma_r(r)$ are the axial, hoop and radial stresses at some distance from the ends of the rod in the region of plane strain. Later, we will quantify the term "some distance." If a cut is made on the plane $z = 0$ in the plane strain region the change of the stresses in the part of length λ shown in Fig. 8.3-b may be obtained by applying stresses on the plane $z = 0$ as shown in Fig. 8.3-b which are equal and opposite to $\sigma_z(r)$. To distinguish these stresses from the initial stress $\sigma_z(r)$ in plane strain we refer to them as $\sigma_z'(r)$, where $\sigma_z'(r) = -\sigma_z(r)$. The stress state produced by $\sigma_z'(r)$ is not one of plane strain but rather, as will be shown, becomes negligible when z is greater than about one diameter. Since $\sigma_z(r)$ is given by Eq. (8.16), with a change of sign, finite element computation may be used to obtain the normal stress $\sigma_z(r, z)$ and the shear stress $\tau_{rz}(r, z)$ corresponding to each term in the series of functions $S_i(r)$ with the amplitude coefficients A_z^i being taken as unity. We use the notation $S_i(r, z)$ and $T_i(r, z)$ to represent the values of $\sigma_z(r, z)$ and $\tau_{rz}(r, z)$ which are obtained by finite element computation when the stresses $-S_i(r)$ are imposed on the face $z = 0$. Thus,

$$\sigma_z'(r, z) = \sum_{i=1}^{n} A_z^i S_i'(r, z) \quad and \quad \tau_{rz}'(r, z) = \sum_{i=1}^{n} A_z^i T_i'(r, z) \qquad (8.18)$$

These expressions are useful in determining the decay of the stresses with increasing z. However, for the single-slice method only the stresses at $z = a$ are required.

Removal of a slice, by a second cut at $z = a$ as shown in Fig. 8.3-c, releases the stresses $\sigma_z(r)$, $\sigma_z'(r, a)$ and $\tau'(r, a)$, which are all functions of $S_i(r)$ and A_z^i. Thus, the strains produced on the face $z = 0$, when the slice is removed at $z = a$, depend only on the functions $S_i(r)$ and the amplitude coefficients A_z^i. For the present procedure we can measure hoop strain, radial strain or both on the face $z = 0$. However, we decided to measure hoop strain since it will be larger than the radial strain near the outside surface ($r = 1$). For a rod with a length to diameter ratio of 6, we used a finite element solution consisting of 1480 quadratic axisymmetric elements to compute the hoop strain $C_i(r)$ at $z = 0$ for each $S_i(r)$ due to the second cut at $z = a$. The computation was carried out for $i = 1$ to 4. Since the loading conditions defined by $S_i(r)$ satisfy equilibrium conditions, a minimum number of constraints were used on the boundary. The hoop strain produced by the entire series of functions of $S_i(r)$

is then given by

$$\epsilon_\theta(r) = \sum_{i=1}^{n} A_z^i C_i(r) \tag{8.19}$$

In the analysis presented so far the strains to be measured on the face of the slice are assumed to be solely due to the release of the residual stresses. However, even under well-controlled testing conditions the measured strain often contains a small "zero shift" error. The effect of this error is equivalent to introducing an additional term, $A_z^0 C_0$, in Eq. (8.19), where C_0 is the strain produced by a uniform axial stress of unit amplitude. The strain measured on the face of the slice now becomes

$$\epsilon_\theta(r) = A_z^0 C_0 + \sum_{i=1}^{n} A_z^i C_i(r) = \sum_{i=0}^{n} A_z^i C_i(r) \tag{8.20}$$

where the term $A_z^0 C_0$ represents the error involved in the strain measurement.

In our experiment using the single-slice method, we found that the inclusion of the error term in Eq. (8.20) improves the estimation of the residual stress by about 5%. The magnitude of A_z^0 relative to the peak value of the measured stress indicates approximately the accuracy of the measurement and computation.

In principle only $n + 1$ measurements of strain are required to obtain the $n+1$ values of A_z^i and determine the axial stress $\sigma_z(r)$. However, to minimize the effect of random error, it is preferable to measure strain at p locations with $p > n + 1$ The problem is now overdetermined, and we use a standard least squares solution to determine the coefficients A_z^i.

For a number p of strain measurements Eq. (8.20) in matrix form may be written as

$$[C]A = \epsilon \tag{8.21}$$

where $[C]$ is a $p \times (n+1)$ matrix with the element at i^{th} row and j^{th} column defined by $C_{ij} = C_i(r_j)$, A is a $(n+1) \times 1$ column vector with the element at the ith row defined by $A_i = A_z^i$ and ϵ is a $p \times 1$ column vector with the element at the jth row defined by $\epsilon_j = \epsilon_j(r_j)$. It may be shown that a least squares fit provides a solution for the amplitude vector

$$\mathbf{A} = \left\{ [\mathbf{C}]^t [\mathbf{C}] \right\}^{-1} \left\{ [\mathbf{C}]^t \epsilon \right\} \tag{8.22}$$

Once the $[C]$ matrix has been obtained, this equation is readily solved using conventional computer programs such as MATLAB or MathCad. After the vector $\mathbf{A}$ is obtained, the first element A_z^0 is discarded and only the amplitude coefficients A_z^i for $1 \le i \le n$ are used in Eq. (8.16) to compute the axial stress.

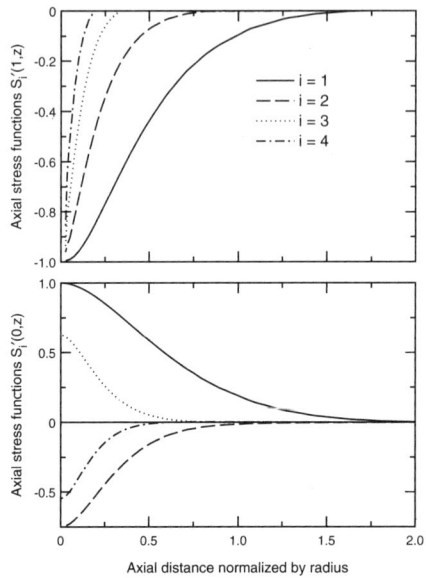

Fig. 8.4. Computed values of $S_i'(r, z)$ produced by loading $S_i(r)$ on the plane $z = 0$. (a) $r = 0$ and (b) $r = 1$.

8.2.3 The Choice of the Functions $S_i(r)$

To implement our solution we need to specify the functions $S_i(r)$. For axisymmetric problems these can be obtained from Eq. (8.15) in Section 8.1.2,

$$\sigma_z(r) = \frac{1}{\nu} \left\{ [\sigma_\theta(r) + \frac{1}{r} \int_0^r \sigma_\theta(r) dr] \right.$$

$$\left. - [\sigma_\theta^s(r) + \frac{1}{r} \int_0^r \sigma_\theta^s(r) dr] \right\} \qquad for\ a\ rod$$

$$\sigma_z(x) = \frac{1}{\nu} \left\{ [\sigma_\theta(x) + \frac{1}{b+x} \int_0^x \sigma_\theta(x) dx] \right.$$

$$\left. - [\sigma_\theta^s(x) + \frac{1}{b+x} \int_0^x \sigma_\theta^s(x) dx] \right\} \qquad for\ a\ cylinder \qquad (8.23)$$

From our past experience in measuring residual hoop stresses in cylinders and rods we have found that expressing the hoop stress as a series of Legendre polynomials has significant advantage compared to say a power series. First, it is more convenient for studying convergence of the solution. Second, if the constant "zero order" Legendre polynomial is excluded, the remaining higher order Legendre polynomials always satisfy the equilibrium conditions which are necessary for computation of the hoop stress, and for the present case, as we will see, also allow the hoop strains $C_i(r)$ in Eq. (8.19) to be computed.

For a solid rod we use the Legendre polynomials as conventionally defined over the range -1 to +1. Because of symmetry the computations are carried out over the range $0 \le r \le 1$, and we choose only the even polynomials $L_{2i}(r)$, $i > 0$ to satisfy equilibrium and ensue zero slope of the stress distribution at $r = 0$.

Thus, we express the hoop stresses in plane strain and plane stress by the summations

$$\sigma_\theta(r) = \sum_{i=1}^{n} A_\theta^i L_{2i}(r) \quad and \quad \sigma_\theta^s(r) = \sum_{i=1}^{n} A_\theta^{si} L_{2i}(r) \tag{8.24}$$

in which the amplitude coefficients A_θ^i and A_θ^{si} have to be determined from an experiment. For a cylinder to obtain the range of the polynomials from -1 to $+1$ we choose $L_i(2x - 1)$. Substituting Eq. (8.24) into Eq. (8.23) provides an expression for the axial stress for a rod

$$\sigma_z(r) = \frac{1}{\nu} \sum_{i=1}^{n} (A_\theta^i - A_\theta^{si}) S_i(r) = \sum_{i=1}^{n} A_z^i S_i(r)$$

$$with \quad S_i(r) = L_{2i}(r) + \frac{1}{r} \int_0^r L_{2i}(r) dr \tag{8.25}$$

or a cylinder

$$\sigma_z(x) = \frac{1}{\nu} \sum_{i=1}^{n} (A_\theta^i - A_\theta^{si}) S_i(x) = \sum_{i=1}^{n} A_z^i S_i(x)$$

$$with \quad S_i(x) = L_i(2x - 1) + \frac{1}{b+x} \int_0^x L_i(2x - 1) dx \tag{8.26}$$

It can be checked that these expressions for $S_i(r)$ and $S_i(x)$ satisfy axial equilibrium which must be the case since they are obtained using hoop stresses which satisfy equilibrium and Eq. (8.23). It may be shown that for a rod $S_i(1) = 1$ for all values of i.

8.2.4 Determination of the Hoop and Radial Stresses in Plane Strain

A useful feature of the single-slice method is that the hoop stresses measured in the slice can be combined with the axial stress to obtain the hoop and radial stresses in the original cylindrical member. From Eq. (8.25) or Eq. (8.26) we may deduce that the relation between the amplitude factors for the axial and hoop stresses in the cylinder and the hoop stress in the slice is given by

$$A_\theta^i = A_\theta^{si} + \nu A_z^i \tag{8.27}$$

To determine the amplitude factors A_θ^{si} for the slice shown in Fig. 8.3-c a single strain gage is attached to the surface in the hoop direction. A cut of progressively increasing depth is made, starting diametrically opposite the strain gage as shown in Fig. 8.3-c. From measurement of strain as a function of the depth of cut, the amplitude factors may be deduced using the slitting method. The compliance function for a solid slice has been given in Section 4.3 and that for a cylinder in Section 4.5.

Substitution of the amplitude factors A_θ^i determined from Eq. (8.27) into the first expression in Eq. (8.24) gives the residual hoop stress in the original cylinder. The radial stress may then be obtained from Eq. (8.3).

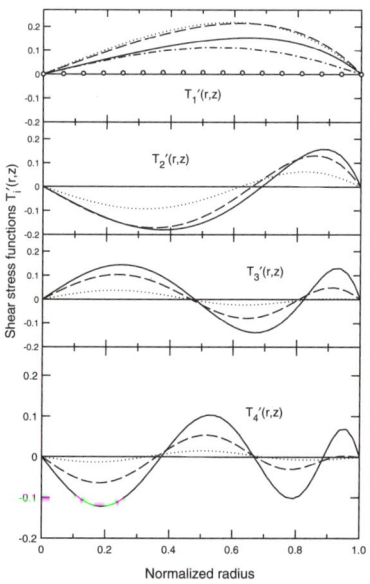

Fig. 8.5. Computed values of $T_i(r, z)$ produced by loading $S_i(r)$ on the plane $z = 0$. Solid lines $z = 0.129$, dashed lines $z = 0.258$, dotted lines $z = 0.45$, dash-dot lines $z = 0.9$, and opened circles $z = 1.8$.

8.2.5 Plane Strain and the Choice of the Slice Thickness

Relatively few discussions of problems involving plane strain provide information on the distance from the ends of a part at which the plane strain assumption is a good approximation. Figure 8.4 shows the first four terms in the series $S_i'(r, z)$ for $r = 0$ and $r = 1$ as a function of dimensionless axial distance z. Higher order terms decay rapidly and even the lowest order term essentially vanishes at one diameter from the plane $z = 0$. Figure 8.5 shows the shear stress functions $T_i(r, z)$. They are zero at $z = 0$, increase at first with z and then decay with increasing z. The lowest order term increases with z up to a value of about 0.5 and then decreases rapidly with increasing z. The higher order terms are seen to decrease more rapidly than the lowest order term. Figure 8.6 shows the hoop stress at $r = 0$ corresponding to $S_i(r)$. It is seen to decay even more rapidly than the axial stress. Thus, it appears that at about one diameter from the ends of the rod, the stress state should be essentially one of plane strain.

From Figs. 8.4-8.6 it is seen that the numerical computation of $S_i(r, z)$ and $T_i(r, z)$ from the function $S_i(r)$ applied at $z = 0$ can be carried out for a length of rod of one diameter rather than using the entire length λ shown in Fig. 8.2-b. Figures 8.5 and 8.6 also provide guidance in the choice of the slice thickness a. In order that the first four terms in the series of functions S_i and T_i ($i = 1$ to 4) contribute to the solution, the value of a should be about one quarter of the radius. For the present study the solution converged using the first three terms. If terms beyond $i = 4$ are required, a more detailed analysis of the strains C_i as function of slice thickness a would be required.

An extension of the experimental and computational procedure may be of value in some applications. If a short section of length λ', shown in Fig. 8.7, is removed from the region which is initially in plane strain, the previous derivation can be applied with a simple modification. Referring to Fig. 8.7, the stresses σ_z and σ_z' have to be applied to both faces of the section of length 1 to calculate the stresses τ_{rz} and σ_z on the plane $z = a$. The values of S_i and T_i are changed, but the derivation follows that given earlier. This approach is clearly of great advantage in dealing with parts of large diameter. The dimension λ' shown in Fig. 8.7 may be as short as the radius of the rod.

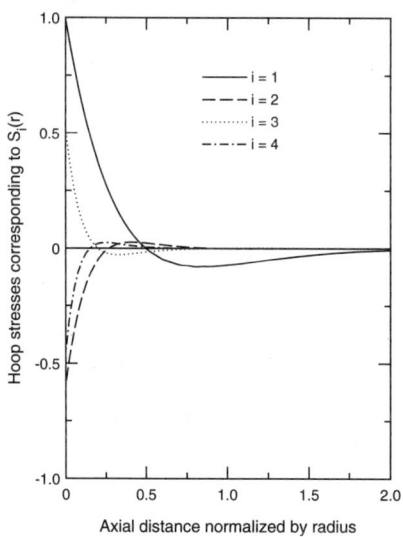

Fig. 8.6. Computed hoop stress at $r = 0$ produced by loading $S_i(r)$ on the plane $z = 0$.

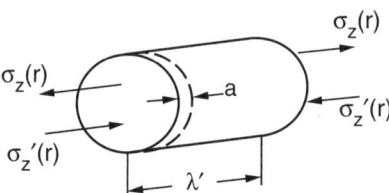

Fig. 8.7. Single-slice method applied to a short section ($\lambda' < \lambda$) cut from plane strain region of the original rod.

8.2.6 An Additional Experimental Feature

Before turning to experimental validation of the preceding derivation, we discuss an optional procedure which should lead to a more precise measurement for the surface stress unless a very steep near surface stress gradient is involved.

If an axial and a hoop strain gage is mounted on the curved surface at the location of the future slice before any cuts are made, then the strains ϵ_z, and ϵ_θ, due to removal of the slice provide a direct, and we assume exact, computation of the axial stress at $r = 1$. That is,

$$\sigma_z(r = 1) = E(\epsilon_z + \nu\epsilon_\theta)/(1 - \nu^2) \tag{8.28}$$

From Eq. (8.16) this leads to

$$\sigma_z(r = 1) = \sum_{i=1}^{n} A_z^i S_i(1) = \sum_{i=1}^{n} A_z^i \tag{8.29}$$

which may be used to eliminate one of the A_z^i from the least squares fit. Since the result is not influenced by the choice of i, for simplicity in both derivation and computation we choose $i = n$, i.e.,

$$A_z^n = \left[\sigma_z(1) - \sum_{i=1}^{n-1} A_z^i\right] \tag{8.30}$$

We now make use of the axial stress at $r = 1$ which we assume is obtained exactly from the hoop and axial strains at the surface. Substituting Eq. (8.30) into Eq. (8.20) leads to

$$\epsilon_\theta(r) - \sigma_z(1)C_n(r) = A_z^0 C_0 + \sum_{i=1}^{n-1} A_z^i \left[C_i(r) - C_n(r) \right] \qquad (8.31)$$

For a number of p strain measurements, Eq. (8.31) in a matrix form becomes

$$\boxed{C}\ \overline{A} = \overline{\epsilon} \qquad (8.32)$$

which, using the notation $C_{ij} = C_i(r_j)$ for $1 \leq j \leq p$ where j describes the radial locations at which C_{ij} is computed,

$$[C] = \begin{bmatrix} C_0 & C_{11} - C_{n1} & \dots & C_{i1} - C_{n1} & \dots & C_{(n-1)1} - C_{n1} \\ & \dots & & \dots & & \\ C_0 & C_{1j} - C_{nj} & \dots & C_{ij} - C_{nj} & \dots & C_{(n-1)j} - C_{nj} \\ & \dots & & \dots & & \\ C_0 & C_{1p} - C_{np} & \dots & C_{ip} - C_{np} & \dots & C_{(n-1)p} - C_{np} \end{bmatrix} \qquad (8.33)$$

and

$$\overline{A} = \begin{bmatrix} A_0 \\ A_z^1 \\ \dots \\ A_z^i \\ \dots \\ A_z^{n-1} \end{bmatrix} \quad and \quad \overline{\epsilon} = \begin{bmatrix} \epsilon_1 - \sigma_z(1)C_{n1} \\ \dots \\ \epsilon_j - \sigma_z(1)C_{nj} \\ \dots \\ \epsilon_p - \sigma_z(1)C_{np} \end{bmatrix} \qquad (8.34)$$

The solution of Eq. (8.32) using a least squares fit has the same form as Eq. (8.22).

8.2.7 Experimental Validation

To produce an axisymmetric residual stress field an apparatus shown schematically in Fig. 8.8 was constructed. An aluminum alloy rod 5 cm in diameter and 38 cm long was heated to about $500°C$ for 30 minutes by electric heating tapes. The tapes and insulation were removed rapidly, and the rod which had 6 mm diameter shafts at each end was transferred to the apparatus in which it was rotated by an electric drill. It was then quenched by water jets from 1.5 mm holes spaced 6 mm apart along the entire length of two pipes located on opposite sides of the rod as shown in Fig. 8.8.

Next, a pair of strain gages were attached in hoop and axial directions on the outer surface in the mid-section and the water-quenched bar was separated, as shown in Fig. 8.3, into three parts: a disk about 6.3 mm thick containing the strain gages on the outer surface, and two cylinders of about equal length. From the change of the strains measured before and after cutting

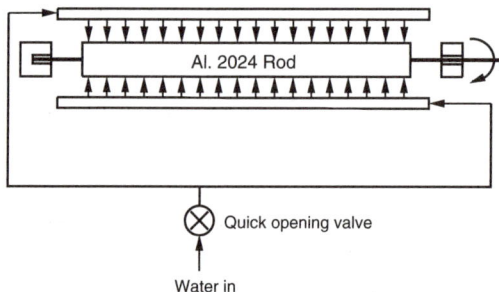

Fig. 8.8. Schematic of the apparatus used to produce axisymmetric residual stresses and plane strain in the central section of the specimen.

out the disk, the residual axial stress on the outer surface was determined to be 54.4 MPa using $E = 68.9$ GPa and $\nu = 0.33$.

The disk and one of the cylinders were used to measure the residual hoop stresses in plane stress and plane strain by the crack compliance method using axial cuts [15]. Strain gages mounted on the surface were used to measure the change in hoop strains as a diametral cut of progressively increasing depth was introduced opposite the strain gages by electrical discharge wire machining (wire EDM), as shown in Fig. 8.3. The residual hoop stresses in the disk and the cylinder were then estimated using even order Legendre series, which were found to converge when the order of the series was equal to 6. Finally, the residual axial stress distribution was obtained using Eq. (8.15). Figure 8.9 shows the distributions of the residual hoop and axial stresses as a function of the normalized radius r. The magnitude of the stresses is seen to be moderate and therefore should provide a pertinent benchmark for checking the sensitivity of the single-slice method.

Turning to the single-slice method, eleven strain gages were attached along a diameter in the hoop direction on the end face of the other cylinder, near the mid-section of the original bar. A 6.3 mm thick slice was then cut out by wire EDM. Hoop strains with a peak value of over $720\mu\epsilon$ were recorded.

Using the axial stress measured on the surface from cutting out the disk, the least squares fit solution of Eq. (8.32) for $n = 3$ was used to obtain the amplitude coefficients A_z^0, A_z^1, and A_z^2 in Eq. (8.20). The amplitude factor A_z^3 was then obtained from Eq. (8.30). Since A_z^0 corresponds to a uniform stress, which is only due to the presence of error in experiment and numerical computation, its value is an approximate indication of the accuracy of the test and computation. In the present test A_z^0 was found to be about 5% of the peak value of the measured axial stress.

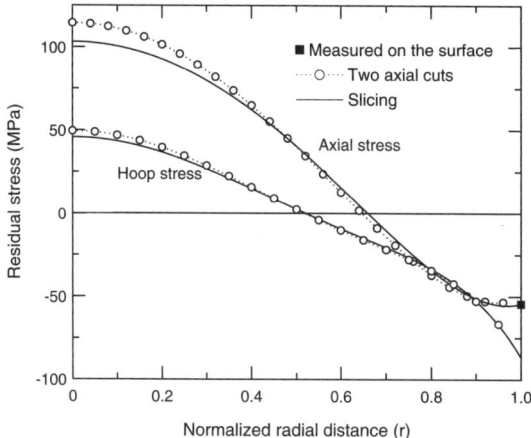

Fig. 8.9. Residual axial and hoop stresses measured by the two-axial-cuts and single-slice methods for a water-quenched rod.

The residual hoop stress measured earlier for the disk was combined with the result for the residual axial stress to obtain the residual hoop stress in the rod using Eqs. (8.24) and (8.27). The residual axial and hoop stresses obtained by the single-slice method are shown as solid lines in Fig. 8.9. It is seen that the overall agreement with the two axial cuts method is very good. It is noted that the two-axial-cuts method is limited to measuring residual stresses up to about 3% from the surface while the single-slice method is capable of measuring residual stress over the entire radius.

8.2.8 Discussion

An approach using a single slice cut out from the mid-section of a specimen to measure a complete 3-D residual stress distribution has been presented. Experimental validation was carried out by measuring the residual stresses in a water-quenched cylinder using the single-slice method and the two-axial-cuts method. The method was found to have excellent sensitivity and led to an accurate result over the entire radius.

Residual stress measurement is often required when a part fails due to fracture under low external loading. Measurements are almost always carried out on a similar uncracked part because the residual stresses in the original part had been altered due to cracking. However, in many situations it is very difficult, if not impossible, to duplicate the residual stress in the original part. The single-slice method provides a solution for this problem if fracture is dominated by elastic deformation because the first step of the method is to cut the part apart. Thus, if the plane of the first cut is chosen such that it is close to the plane of fracture, the original normal stress distribution can be measured by cutting out a slice near the plane of the fracture.

Another important feature of the present method is that the requirement on the length of the specimen is much less stringent than those required by other methods. A slice can be cut out from a specimen of length less than that required by the condition of plane strain as long as the specimen is first cut out from the plane region of a long part. The only difference between using a long specimen and a short one is that the sensitivity for the latter is lower. To demonstrate this feature, we cut out a section of length equal to the radius from the central region of the aluminum rod before attaching hoop strain gages and taking a slice. No surface axial stress measurement could be made, but the outer hoop strain was measured very close to the surface. The stresses computed were within a few percent of those shown in Fig. 8.9.

9

Estimation Using Initial Strains

9.1 Introduction

Residual stresses in many long structural parts can be idealized as distributed uniformly in the axial direction with little out-of-plane shear stresses. Examples are rods, beams, rails, and long butt or fillet welds between two plates. In many cases the residual stresses arise under conditions which are not easily simulated by numerical computation. Figure 9.1-A shows schematically a section of rod and beam while Fig. 9.1-B shows a section of butt weld between two plates. In these examples the constraint is often such that residual stress in the z-direction would be expected to be the most severe. Therefore, an essential aspect in assessing the integrity of these parts is to measure the distribution of axial (z-axis) stress.

To obtain the residual stresses, the unknown quantities to be estimated usually correspond to the stresses or average forces released by removing or separating material. Thus, it is commonly believed that the residual stresses are "lost" if any deformation due to partial relief of the stresses, such as fracture of a part, is not sufficiently recorded. This is true for most methods based on an approximation of stresses. One exception is the single-slice method [29] presented in Section 8.2 for axisymmetric bodies for which the original residual stress may be estimated from a section removed from the original body. However, for an arbitrary prismatic body the analysis based on an approximation of the stress becomes much more complicated, and it is not easy to find a series of functions that always satisfy the equilibrium condition.

To overcome these limitations, we turn to an alternative approach based on the approximation of the incompatible initial strain field from which the residual stresses may be obtained. The relation between residual stresses and initial strains has been addressed in many classic textbooks on elasticity (Fung [56], Timoshenko and Goodier [124]). However, its use for residual stress measurement is relatively new. Ueda and his colleagues first proposed their "inherent-strain" method in the mid-70's. Instead of solving the stresses alone, the inherent-strain method aims at establishing the incompatible initial strain

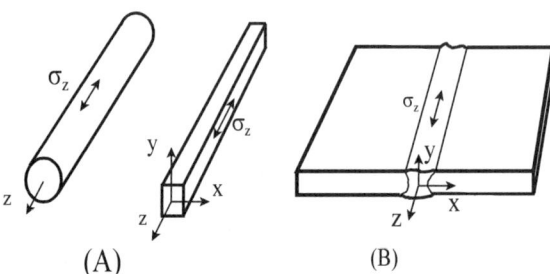

Fig. 9.1. Illustrations of (A) a beam and a rod and (B) a long butt weld between two plates with residual axial stresses distributed uniformly along the length except near the ends.

field that causes the residual stresses. This involves measuring the deformation in different directions due to cutting and/or sectioning the body into strips and many small segments (Ueda et al. [128], Ueda and Fukuda [129], Ueda et al. [127]). With elaborate efforts in both experiment and computation, they demonstrated that inherent strains can be successfully used to estimate residual stresses. The inherent-strain method was later modified by Hill and Nelson [66, 67, 68] to reduce the region required for strain measurement. The modified approach is referred to as the eigenstrain method.

In the present approach [30] initial strains are introduced only for generating a series of stress distributions on the plane of measurement that are used to approximate the axial stresses released prior to and/or during measurement. For this reason it is informative to mention some distinctive features of the approach.

First, a residual axial stress field is uniquely determined by a given incompatible initial strain field but the reverse is not true. This implies that there exists more than one incompatible initial strain field that leads to the same axial residual stress distribution on the plane of measurement.

Second, if the residual stress is altered by cuts made on planes on which the out-of-plane shear stresses are small enough to be neglected before cutting, the original residual axial stress may be estimated using the initial strain field measured from a section of the part as long as little permanent deformation occurs during cutting. Thus, the initial strain field estimated from a fractured part may be used to compute the original normal stress on the plane before fracture if the original residual shear original stress on the same plane was negligible.

Third, for residual stresses produced by local incompatible deformation, such as due to welding, it is only necessary to define the initial strains in a subregion containing the weld (Ueda et al. [127]). This feature makes the approximation of the initial strains due to welding easier than the approximation of the stress distribution over the entire region of the part. It also implies that the original residual stress distribution can be estimated only if the specimen contains the entire initial strain field. For a welded part, for instance, it means that the cross-section of the specimen should be large enough to contain the entire cross-section of the weld.

It is worth noting that the first feature mentioned above is unique to the present approach and, as demonstrated later, it makes the construction of the initial strain distributions for numerical computation relatively easier than the "inherent-strain" approach.

In this section an analysis based on the approximation of initial strains is incorporated into the slitting method and the single-slice method described in this book. A computational procedure is then presented to obtain the change of strains due to stresses released for each given initial strain function. After the initial strain field is determined by a least squares fit over the experimentally measured strains, the residual axial stress can be computed.

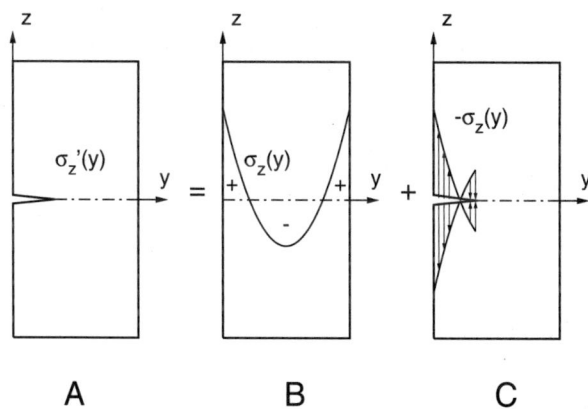

Fig. 9.2. Linear superposition used for crack compliance method based on an approximation of the stress.

To demonstrate the use of initial strains for residual stress measurement, an experiment using the crack compliance method was carried out on a beam which had been cut apart in a previous test. The original stress in the beam before it was cut apart was then estimated using the initial strain approach. It is shown that the result agrees very closely with the analytical solution.

9.2 Initial Strain Approach for the Crack Compliance Method: Axial Stress in a Beam

The crack compliance method has been used to measure residual stresses through-the-thickness in parts of different configurations [52, 98]. In this method the residual stress distribution to be measured is approximated by a series of continuous or piecewise functions with n unknown coefficients. From the linear superposition shown in Fig. 9.2, the deformation due to introducing a cut of progressively increasing depth is the same as that produced by applying the stress being released on the faces of the cut with the sign reversed. This allows the deformation or the "compliance function" due to each individual function in the approximate stress distribution to be obtained. Since the number m of measurements recorded during cutting is usually much larger than the number n of unknown coefficients, a least squares fit is commonly used to determine the unknowns.

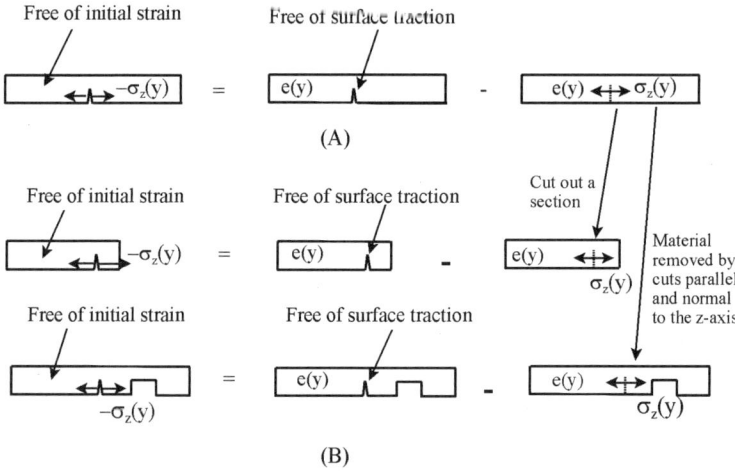

Fig. 9.3. Schematics of (A) crack compliance method for a beam subjected to initial strains with and without a cut of increasing depth; (B) after undergoing cutting or material removal the same beam is used to measure the initial strains by introducing a cut of increasing depth.

To demonstrate how the initial strain approach works for the crack compliance method, we consider the residual stress produced by a beam subjected to pure bending beyond its elastic limit as shown in Fig. 7.1. Since the residual stress is uniform along the length between the two inner support pins, it is sufficient to assume that the initial strain field is uniform along the entire length of the beam. Also, since there exists more than one initial strain field which produces the same axial stress in the beam, we choose the one that corresponds to the deformation caused by a temperature variation $e(y)$, which, when expressed in terms of a series of Lengedre polynomials $L_i(y)$ [70], leads to

$$e(y) = e_x(y) = e_y(y) = e_z(y) = \sum_{i=2}^{n} A_i L_i(y) \quad for \ -1 \leq y \leq 1 \quad (9.1)$$

in which y is the normalized distance measured from the neutral axis. It is seen that the uniform and linear terms are omitted from Eq. (9.1) because they do not produce any residual stress for a beam free of any constraint. The stress in the axial (z) direction can readily be obtained using the analytical solution available for thermal stresses in a long strip [124]. That is,

$$\sigma_z(y) = -Ee(y) + \frac{E}{2} \int_{-1}^{1} e(y)dy + \frac{3Ey}{2} \int_{-1}^{1} e(y)y\,dy \quad (9.2)$$

in which E is the elastic modulus. Since the integrals in Eq. (9.2) vanish for all Legendre polynomials with $i \geq 2$, we arrive at a very simple solution for the stress,

$$\sigma_z(y) = -Ee(y) = -E\sum_{i=2}^{n} A_i L_i(y) \quad for \ -1 \leq y \leq 1 \quad (9.3)$$

Now we introduce a cut of increasing depth to the beam. The deformation due to the release of the stress on the plane of cut can be obtained, as shown in Fig. 9.3-A. Thus, without specifying explicitly the stress distribution on the faces of the cut, the deformation due to releasing the stress by cutting can be determined from the difference between the bodies with and without the cut. This analysis can be generalized to other situations, as shown in Fig. 9.3-B, in which the shape of the beam is altered by material removal without any plastic deformation. Although the stresses in latter cases are different from the original ones, the deformation due to introduction of a cut of increasing depth is caused by the same initial strain or axial stress. Therefore, the original stress can still be computed directly from Eq. (9.3) once the initial strain field is estimated. On the other hand, if the existing stresses on the plane of cut are to be measured, they need to be computed using the estimated initial strains for the exact geometry of the part shown in Fig. 9.3-B without the cut.

Denoting the strains produced by the initial strains with and without the cut by ϵ_c and ϵ_o respectively, the change in strain due to cutting is then given by

$$\epsilon(a_j) = \epsilon_c(a_j) - \epsilon_o = \sum_{i=2}^{n} A_i[\epsilon_{ci}(a_j) - \epsilon_i^o] = \sum_{i=2}^{n} A_i \epsilon_i^e(a_j) \qquad (9.4)$$

where a_j is the jth depth of cut and ϵ_i^e is the strain or compliance function produced by the ith order function $L_i(y)$ in Eq. (9.1). For a number m of depths Eq. (9.4) can be written in a matrix form

$$[\epsilon^e] \ \mathbf{A} = \epsilon \qquad (9.5)$$

where

$$[\epsilon^e] = \begin{bmatrix} \epsilon_{21}^e & \cdots & \epsilon_{i1}^e & \cdots & \epsilon_{n1}^e \\ \cdots & \cdots & \cdots & & \\ \epsilon_{2j}^e & \cdots & \epsilon_{ij}^e & \cdots & \epsilon_{nj}^e \\ \cdots & \cdots & \cdots & & \\ \epsilon_{2m}^e & \cdots & \epsilon_{im}^e & \cdots & \epsilon_{nm}^e \end{bmatrix} \cdot \mathbf{A} = \begin{bmatrix} A_2 \\ \cdots \\ A_i \\ \cdots \\ A_n \end{bmatrix} \quad and \quad \epsilon = \begin{bmatrix} \epsilon_1 \\ \cdots \\ \epsilon_j \\ \cdots \\ \epsilon_m \end{bmatrix} \qquad (9.6)$$

For $m > n$ a least squares fit can be used to obtain the unknown coefficient vector $\mathbf{A}$ from the measured strain vector ϵ. This leads to

$$\mathbf{A} = \left\{ [\epsilon^e]^t [\epsilon^e] \right\}^{-1} [\epsilon^e]^t \epsilon \qquad (9.7)$$

where a superscript t denotes a transposed matrix. In practice, to reduce the influence of experimental error and establish the optimum order of estimation, an estimation procedure such as one described in Chapter 7 should be used to solve for the unknown coefficients. Substituting the computed coefficients A_i into Eq. (9.3), the original residual stress distribution can be obtained.

9.3 Initial Strains Approach for the Single-Slice Method: Axial Stress in a Rod

The single-slice method [29], described in Section 8.2, is primarily for measuring the axial stress distribution in the mid-section of a part. Experimentally, it involves measuring the strain due to removing a slice from the mid-section. First, a complete cut is made to separate the part in the mid-section. On one of two sections strain gages are installed on the surface exposed by the first cut, and a second cut is made to remove a slice containing the strain gages while the change of strains due to removing the slice is recorded. The axial stress variation over the cross-section can then be estimated using the strain

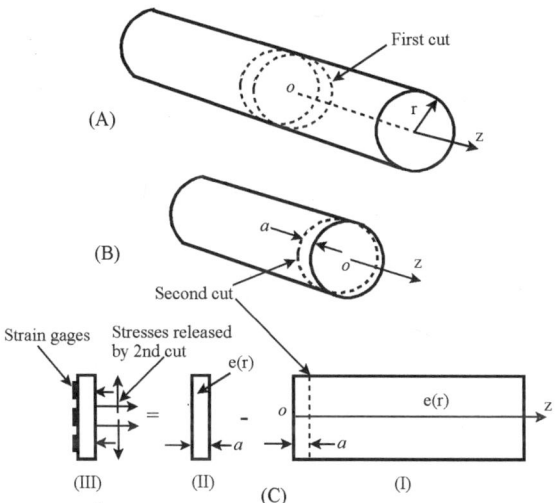

Fig. 9.4. Schematics of (A) a long cylinder separated at the mid-section $z = 0$; (B) a slice cut out along plane $z = a$ while hoop and/or radial strains are recorded on plane $z = 0$; (C) linear superposition for the initial strain approach.

data.

Because the analysis described in Section 8.2 for axisymmetric problems is based on the approximation of the stress, a rigorous solution is given here to show that the initial strain approach provides an improved alternative. Consider a rod in plane strain with an axisymmetric residual stress field, shown in Fig. 9.4-A. As mentioned earlier, the choice of the initial strains is not unique. To illustrate this point, an initial strain field which is uniform in the z-direction is chosen as

$$e_z(r) = e(r)$$
$$e_r(r) = e_\theta(r) = \beta e(r) \tag{9.8}$$

where r is the normalized radial distance and β is an arbitrary constant. Obviously, a variety of initial strain fields may be produced by using different values of β. In this case the analytical solution for the corresponding stress distribution in the z-direction may be obtained as

$$\sigma_z(r) = \frac{E(\beta\nu + 1)}{1 - \nu^2} \left(2 \int_0^1 erdr - e \right) \tag{9.9}$$

where E is the elastic modulus and ν the Poisson's ratio. It is seen from

Eq. (9.9) that any initial strains defined by Eq. (9.8) lead to similar distributions of the axial stress even though they produce different stresses in the r and θ-directions. Thus, they all can be used to estimate the same residual axial stress because the deformation measured by cutting off the slice is only dependent on the release of the axial stress. Particularly, the axial stress can always be generated by specifying an initial strain in the axial direction.

For axisymmetric problems, a complete polynomial series consists of only even orders $2i$ and the initial strain e may be expressed as

$$e(r) = \sum_{i=1}^{n} A_i \, r^{2i} \tag{9.10}$$

in which A_i is the amplitude coefficient for the $2i^{th}$ order term. From Eq. (9.9) the residual axial stress produced by the initial strain given in Eq. (9.8) becomes

$$\sigma_z(r) = \frac{E(\beta\nu + 1)}{1 - \nu^2} \sum_{i=1}^{n} A_i \left(\frac{1}{i+1} - r^{2i} \right) \tag{9.11}$$

It can readily be shown that equilibrium condition is always satisfied by Eq. (9.11). Since Eq. (9.10) is defined by a complete even order polynomial series, the corresponding axial stress distribution is also a complete even order polynomial series. Therefore, the axial stress distribution obtained earlier in Eq. (8.23) can be constructed exactly by the polynomials given by Eq. (9.11).

Now the strain or deformation produced by each term of the stress given in Eq. (9.11) can be computed using each term of the initial strains given in Eq. (9.10). Setting $\beta = 1$, the assumed initial strain field becomes one produced by a temperature distribution, which can be handled directly by most finite element programs. Setting $\beta = 0$, on the other hand, the only non-zero initial strain is in the axial direction, which can be simulated by specifying a temperature distribution while setting thermal expansion coefficients in the r and β directions to zero.

We now separate the rod shown in Fig. 9.4-A by a complete cut in the mid-section. For elastic deformation the initial strain given by Eq. (9.10) remains unchanged, and the new stress state for either half of the body can be determined from the same initial strain field. We now introduce another complete cut on plane $z = a$ to remove a slice of thickness a as shown in Fig. 9.4-B. The change of strain on the surface of the slice exposed by the first cut is due to release of the stresses on the plane $z = a$. If the approach based on the approximation of the stress were used, the normal and shear stress distributions on the plane $z = a$ before the second cut would have to be computed first. Then, with a reversed sign, they would be used as the loading conditions on the face exposed by the second cut shown in Fig. 9.4-B to compute the deformation of the slice due to the second cut. For the approach based on initial strains, however, we do not need to compute the normal and shear stresses existing on plane $z = a$, nor do we need to specify any additional loading conditions other than the initial strains in the slice. A simple procedure based on linear superposition shown in Fig. 9.4-C may be used to obtain the deformation due to cutting out the slice. It is seen that the deformation produced by loading on a slice of thickness a without any initial strain, shown as (III), is equal to the difference in the deformation obtained for the slice, shown as (II), subjected to only initial strain and the deformation for one half of the rod, shown as (I), subjected to only the same initial strain. Thus, the change of strain on the face of the slice, $z = 0$, due to cutting off the slice can be obtained for each function given in Eq. (9.10) without knowing the stresses on plane $z = a$. This leads to a more straightforward implementation than that based on the approximation of the stress.

Using the linear superposition approach shown in Fig. 9.4-C, the change in strain, say in the θ-direction, due to cutting off the slice may be expressed as

$$\epsilon_\theta(r_j) = \epsilon_\theta^c(r_j) - \epsilon_\theta^o(r_j) = \sum_{i=1}^n A_i[\epsilon_{\theta i}^c(r_j) - \epsilon_{\theta i}o(r_j)] = \sum_{i=1}^n A_i \epsilon_{\theta i}^e(r_j) \quad (9.12)$$

where r_j is the radial location of the jth strain gage and $\epsilon_{\theta i}^e$ is the strain produced by the ith order function r^{2i} in Eq. (9.10). For a number m of strain gages Eq. (9.12) can be written in a matrix form

$$[\epsilon_\theta^e] \; \mathbf{A} \; = \; \epsilon_\theta \quad\quad\quad (9.13)$$

where matrix $[\epsilon_\theta^e]$, column vectors $\mathbf{A}$ and ϵ_θ have the same forms as the counterparts in Eq. (9.6). Similarly, for $m > n$, the unknown coefficient vector $\mathbf{A}$ can be obtained by a least squares fit as given by Eq. (9.7).

9.4 Experimental Validation

The residual stress produced by four-point bending, as illustrated in Fig. 7.1, has been used to validate the crack compliance method in the past [101]. This is because the residual stress can be predicted accurately using the stress-strain relation obtained from the same beam [79]. Also, it provides an ideal benchmark for checking the capability of the method for measuring a rapidly varying residual stress field through the thickness since residual stresses with different magnitude and gradient can be generated by controlling the extent of the plastic deformation. As demonstrated Chapter 7, a residual stress distribution was measured successfully by the crack compliance method using the compliance functions computed either by LEFM solutions or numerically by finite element method. In this section we present a more severe test. A residual stress distribution was generated in another beam, which had a steeper gradient near the mid-plane of $y/t = 0.5$. The material properties and test configuration are given in Table 9.1. Because of different yield stresses in compression and tension, the residual stress distributions produced by bending is not exactly antisymmetric, and the peak tensile stress is always a few percent higher than the peak compressive stress. As a comparison, the two stress distributions estimated analytically from two bending tests denoted as A and B are shown in Fig. 9.5. The experimental results using crack compliance method for beam A are also reproduced in the figure. It is seen that the stress gradient in beam B changes much more abruptly near the peak stress than that in beam A.

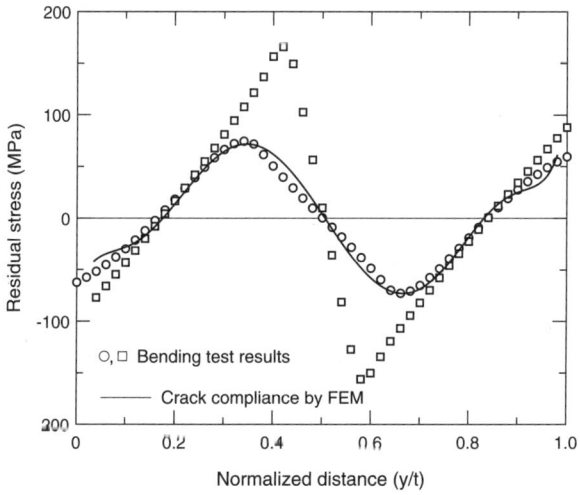

Fig. 9.5. Two different residual stresses (○ - Beam A and □ - Beam B) produced by four-points bending. Solid line - estimated using the slitting method.

Beam B was subsequently separated by wire EDM near the mid-section. Obviously, the stress had been partially released within a region about one thickness from the end exposed by cutting.

Table 9.1. Configuration of the Measurement

Beam Geometry			Stainless Steel 304L			Measurement Configuration	
Width (mm)	Thickness (mm)	Length (mm)	E (MPa)	ν	Cut width (mm)	Gage location	Gage length (mm)
18.77	18.77	78	196	0.3	0.236	Back face	0.8128

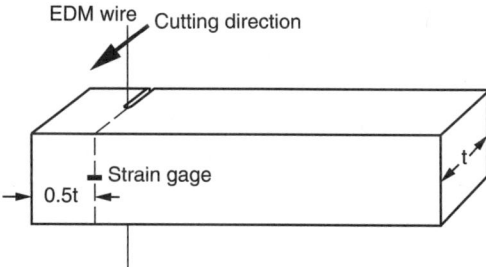

Fig. 9.6. Configuration of the test for measurement of the initial strains in a section of the beam using the slitting method.

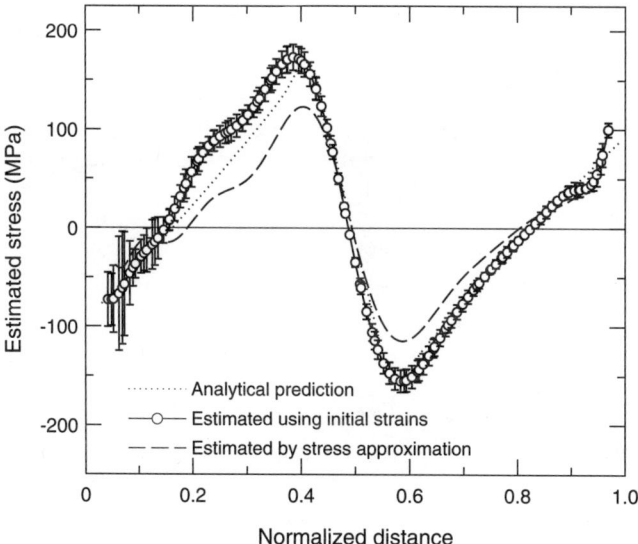

Fig. 9.7. Comparison of the stress distributions estimated using the initial strain approach and the stress based approximation.

In the present study one section of the beam was used to measure the original residual stress by introducing a cut of progressively increasing depth on a plane about one half (51%) thickness away from the end, as shown in Fig. 9.6, while the change in strain was recorded on the back face directly opposite the cut. First, the residual stress on the plane of cut (about half thickness away from the end) was estimated based on approximation of the stress using Legendre series. Convergent results were obtained when the order of approximation was larger than 14. Figure 9.7 shows the estimated stress distribution represented by a seventeenth order polynomial as a dashed line. The estimated peak stress is seen to be about 27% lower than the analytically predicted original residual stress. Next, an initial strain field represented by Eq. (9.1) was used to obtain the compliance functions, as given in Eq. (9.4), and the unknown coefficients are estimated by the procedure described in Chapter 7 using a weighted LSF. The original stress was then computed using Eq. (9.3). Figure 9.7 also compares the stress distributions estimated by the two approaches. It is seen that the initial strain approach gives a correct estimation of the original peak stresses and the measured stress agrees closely with the predicted residual stress. It is worth noting that the measured peak tensile stress is seen to be a few percent higher than the peak compressive stress, which is consistent with the analytically predicted stresses. Also, because the steep stress gradient requires a higher order approximation, the uncertainty is estimated to be substantially larger in the region of $a/t < 0.2$.

In both cases, the crack compliance functions are computed by FEM. In the first case, a fine mesh is required only in the region near the cut, and a coarser mesh is used in the region away from the cut. In the second case, the same fine mesh is used over a larger region, about one thickness of the beam to maintain a uniformly distributed initial strain field along the length of the beam, which requires a considerably larger amount of computation. Also, computations need to be carried out on two configurations for each term of the initial strain, one for the beam with a cut and the other without a cut. Fortunately, the tremendous increase in computing power in the last few years has made the computations feasible for most two-dimensional problems. In fact, all of the computation required for the present measurement was carried out on a personal computer.

9.5 Application: Measurement of the Residual Stress in a Pyrolytic Carbon Coated Graphite Leaflet

Cardiac devices constructed with thin pyrolytic carbon-coated graphite parts are operated in hostile physiologic environments with cyclic loading. Under such conditions, the residual stresses that arise during fabrication may have a significant influence on the fatigue life [110]. In this section we present an application of the initial strain method to the measurement of the residual stresses in thin carbon-coated graphite leaflets.

Typical cross-section view of the leaflets shows that the pyrolytic carbon coating is of the same thickness on both sides of the substrate, Fig. 9.8. Thus, we may assume that the geometry and the variation of the material properties are symmetrical about the mid-plane of the leaflets.

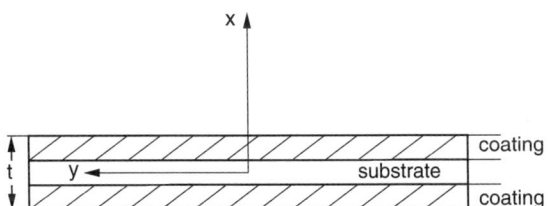

Fig. 9.8. Cross-section view of a pyrolytic carbon coated graphite leaflet.

For through-thickness measurement, a cut is usually approximated by a crack. For a thin leaflet, however, the width as well as the shape of the cut made by a wire EDM must be taken into account in the analysis and computation. For this reason the finite element method is used to model the geometry of a slot with a semicircular bottom. Also, in comparison with the thickness of the leaflets ($t = 0.625$ mm to 0.889 mm), the thickness of the strain gage backing film (0.076 mm) may no longer be negligible and its effect on measured strain needs to be estimated. The corrected strain readings can then be used in estimation of the stress.

Furthermore, the incompatible thermal expansion coefficients across the interface of the graphite substrate and pyrolytic carbon coating are expected to introduce an abrupt change in residual stress, as found later in measured strains when a cut is made across the interface. In this case, instead of estimating the stress directly, we first estimate the initial strain field that causes the residual stress. The initial strain is defined as two independent parts, one that is uniform in the coating and substrate but discontinuous across the interface and another one that is continuous over the entire thickness. The first part represents the difference in the thermal expansion coefficients and the second one represents the continuous variables, such as those due to temperature variation during fabrication.

For a symmetrical configuration the continuous initial strain field $\epsilon^i(x)$ consists of only even order functions, which, for a $2n^{th}$ order approximation, may be expressed as

$$\epsilon^i(x) = \sum_{j=0}^{n} A_j \, x^{2j} \quad for \; -1 \leq x \leq 1 \tag{9.14}$$

where x is the normalized distance measured from the mid-plane and A_j are

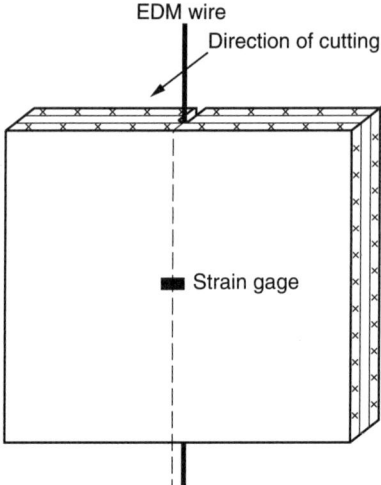

Fig. 9.9. A cut of finite withd introduced by wire EDM through the thickness of a pyrolytic carbon coated graphite leaflet.

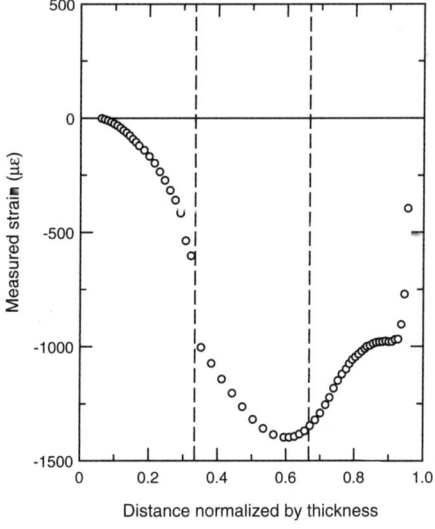

Fig. 9.10. Strain variation measured by the slitting method shown in Fig. 9.9.

the amplitude factors to be determined. Note that the uniform term ($j = 0$) is retained in Eq. (9.14) because of the difference in thermal expansion coefficients. For a number $m > n + 1$ of depths, the unknown coefficients A_j can be estimated by a least squares fit given by Eq. (9.7). The residual stress due to $\epsilon^i(x)$ is then obtained by carrying out another FE computation for a part without the cut.

The configuration of the measurement is shown in Fig. 9.9 and the measured strain is plotted in Fig. 9.10. The approximate locations of the interface are marked by dashed vertical lines. It is seen that the magnitude of readings is significant and discontinuity occurs near the interface between the coating and the graphite substrate.

The next step is to obtain the compliance functions due to initial strains. For a given initial strain field, the residual stress can be determined uniquely, but the reverse is not true. This implies that, in theory, the approach is not necessarily dependent on the choice of the thermal expansion coefficients to be defined in the analysis. Since the thermal expansion coefficients (α_s and α_c for substrate and coating respectively) were not known exactly, different combinations of thermal expansion coefficients of $\alpha_c = 3$ to $5 \times 10^{-6} K^{-1}$ and $\alpha_s = 2$ to $4 \times 10^{-6} K^{-1}$ with $\alpha_c - \alpha_s = 1$ to $3 \times 10^{-6} K^{-1}$ were chosen for definition of the uniform initial strains respectively in the coating and the substrate.

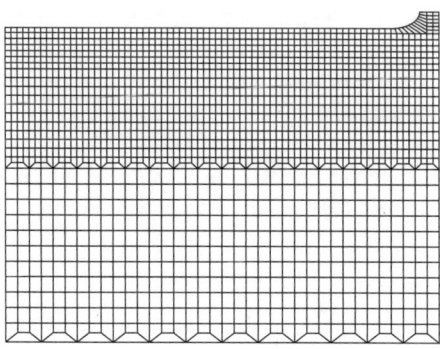

Fig. 9.11. Element mesh used near a cut of finite width and progressively increasing depth.

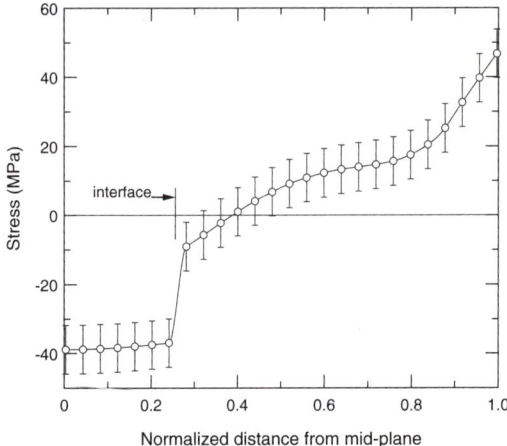

Fig. 9.12. Residual stress distribution measured by the slitting method using the initial strain approximation.

Using the material properties listed in Table 9.2, compliance functions were obtained by FE computation for the initial strain functions defined by Eq. (9.14). More than 3,200 8-node elements are used for one half of the specimen, symmetric about the plane of cut. Figure 9.11 shows a typical element mesh used around the bottom of the cut.

Additional FE computation was also carried out to obtain the correction on the influence of the gage backing thickness on the measured strains. After the initial strain field was obtained by LSF, the residual stress distribution over the entire thickness was obtained by another FE computation with the same thermal expansion coefficients as those used in the computation of the compliance functions. The estimated stresses were found insensitive to the different choices of α_c and α_s.

It is seen from Fig. 9.12 that the estimated stress is compressive in the graphite substrate and tensile in the pyrolytic carbon coating. The peak tensile stress is found near the surface. Below the surface, the stress decreases but remains tensile in the pyrolytic carbon coating. Near the interface, the stress changes rapidly from tensile to compressive. The compressive stress in the substrate is fairly uniform.

As shown in Fig. 9.10, the large magnitude of the measured strain indicates that the compliance method is very sensitive. Although the residual stress in the pyrolytic carbon coating could be measured by placing strain gages near the cut, the very high strain gradient near the cut over the gage length introduces a considerable uncertainty in the presence of a small error in the distance measured from the gage center to the edge of the cut. A similar problem of equal magnitude is expected to arise also for the hole-drilling technique, for which some small misalignment is almost unavoidable. The

Table 9.2. Material properties and configuration used in the stress measurement on a pyrolytic-carbon-coated leaflet

Material	Elastic modulus	Poisson's ratio	Thickness
Substrate	12.74 GPa	0.25	0.28 mm
Coating	28 GPa	0.2	0.28 mm
Strain gage film	3.5 GPa	0.25	0.076 mm

approach presented here is much less influenced by the error in the measured position of the strain gage. Because of the thinness of the leaflet, the material properties of the pyrolytic carbon coating as well as the graphite substrate must be included in the analysis and computation. Otherwise, a significant error would result from assuming a uniform material property.

9.6 Discussion

The "inherent-strain" approach developed by Ueda and his colleagues represents an important and powerful concept that applies to any residual stress measurement. However, a direct implementation of their approach is conceptually nontrivial, and experimentally and computationally expensive. Based on the same concept, we presented an approach that utilizes initial strains for measurement of residual axial stresses. Analytical solutions for the slitting method and the single-slice method were obtained for beams and rods. Experimental validation was carried out by measuring the original residual stresses in a beam subjected to four-points bending and subsequently separated in the mid-section. The residual stress distribution to be measured had a gradient that is considerably steeper than the residual stresses estimated previously by a single continuous function. With the strains recorded when a cut of progressively increasing depth was introduced on a plane about half thickness away from the end, the original stress distribution is successfully estimated and found to agree well with the analytically predicted stress distribution.

The initial strain approach is also shown to provide a simplified analysis for the single-slice method for the measurement of axisymmetric residual axial stresses, such as the water-quenched rod described in Chapter 8.

An interesting application of the initial strain approach is presented for the measurement of residual stress through-thickness of a pyrolytic carbon coated graphite leaflet. The discontinuity in residual stress across the interface between the substrate and coating is captured naturally using two initial strain fields, one for the difference in thermal expansion coefficients and another for the rest of the incompatible deformation. Since the second one is continuous, its estimation is much easier than a direct estimation of a discontinuous stress field.

In summary, the original residual stress can be measured even if the stress has been partially released by cutting as long as the permanent deformation introduced by cutting is negligible and the original stress was uniform over a short distance along the length. This is true for both the slitting method and the single-slice method. The initial strain approach also leads to a more pertinent approximation of the residual stresses that arise from the discontinuity in thermal expansion coefficients even if the actual values of the coefficients are not well defined. The latter application is especially useful when the residual stresses are measured at a temperature different from the operating temperature. With the initial strain fields obtained at two known temperatures, the residual stress in any different operating environment may be predicted so long as the deformation remains linearly elastic.

Residual Stresses and Fracture Mechanics

10.1 Introduction

The measurement of residual stresses in a part and the influence of residual stresses on fracture are related topics which have been studied extensively in the literature. Generally, compressive residual stresses are found to be beneficial in fracture calculations while tensile residual stresses degrade the strength of a part. However, in the first part of this chapter we will point out that the local compressive residual stresses, and sub-surface cracks produced by scratching glass at very low loads are responsible for the low tensile strength of conventional glass specimens. Also, if parts containing surface cracks are exposed to processes, such as shot peening, which induce high, near surface compressive stresses, the internal end of the crack will experience tensile loading. What is often ignored is that the compressive stresses close the crack at the surface and make it more difficult to detect by dye penetrant techniques [83]. An attempt is made here to quantify these observations using procedures based on linear elastic fracture mechanics (LEFM).

The second section of the chapter deals with the measurement of stress intensity factors based on solutions from LEFM. The required computation of a stress intensity factor from a given residual stress is averted. That is, by introducing a thin cut of progressively increasing depth, while measuring strain at a selected location, the stress intensity factor due to the unknown residual stresses may be obtained as a function of cut depth.

10.2 Influence of Residual Stress on Fracture Strength of Glass

Since the pioneering work of Griffith [61] the low strength of glass under tensile stresses has been attributed to an inherent distribution of flaws. Experiments have shown that glass specimens, in both bulk and fiber form, have very high strength, as long as no mechanical contact is made after the surface is formed from the melt [84]. Other observations [47] suggest that accidental mechanical damage to the surface greatly lowers the strength of glass. The concept that inherent flaws exist only at surface in glass appears to be well ingrained in the engineering literature. For example, in the authoritative textbook by McClintock and Argon [80] it is stated, "In some brittle materials, such as inorganic glasses, cracks are formed only at the free surfaces."

To examine this assumption, a simple model which treats the inherent flaws as edge cracks and ignores residual stresses can be used to estimate crack size. Taking a typical value of the fracture stress σ of soda-lime glass in an inert atmosphere of 70 MPa [5] and a fracture toughness K_{Ic} of 0.76 MN/m$^{3/2}$ [131] leads to an estimate of crack size of

$$a = \left(\frac{K_{Ic}}{1.12\sigma}\right)^2 \frac{1}{\pi} = 30\mu m \qquad (10.1)$$

Such a flaw size should be detectable by optical or scanning electron microscopy (SEM). However, to our knowledge no such surface cracks have been observed directly. One explanation is that the crack faces are touching, but our examination by SEM of glass specimens subjected to bending loads which should separate the faces of a surface crack has not revealed cracks. This observation, or rather lack of observation, suggests that the strength impairing flaws may be sub-surface and very close to the surface so that they may be removed by surface etching or melting techniques.

The most likely source of mechanical damage to a glass surface is scratching by a small abrasive particle. This process which is also described as scribing is shown in Fig. 10.1. Above a threshold load a median crack initiates at the bottom of the plastic zone. The lowest threshold load we have observed in experiments is only $0.014N$. Thus, median cracks could well be introduced by wiping a glass surface with a cloth containing small dust particles or in other situations involving mechanical contact.

Since the size of a median crack just above the threshold load is only about several micrometers, its influence on the fracture strength would be very small if no residual stress is present. However, the process of scribing produces a high compressive stress in the plastic zone which prevents the subsurface flaw from growing into the plastic zone. Moreover, the plastic zone also exerts an opening force which leads to a positive stress intensity factor at the lower crack tip shown in Fig. 10.1. The continuing growth of the median crack observed after scribing indicates that the influence of the residual stress is very significant. We have obtained the residual stress distribution below the plastic zone and computed the corresponding stress intensity factor for a subsurface flaw [23]. As shown in Fig. 10.1, the estimated fracture stress of soda-lime glass is greatly reduced by the presence of the compressive residual stress near the surface. These predictions for both inert and moist environments are in good agreement with range of strength values quoted in the literature.

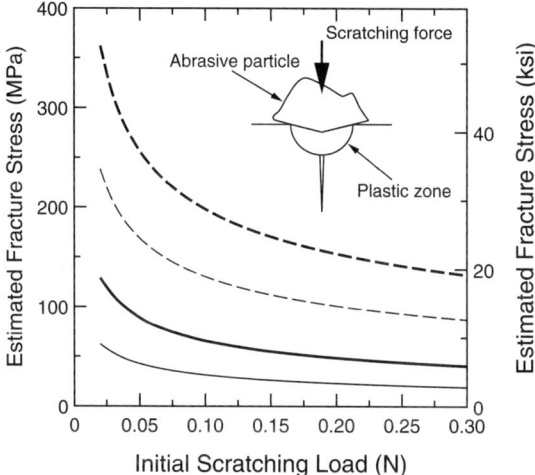

Fig. 10.1. Predictions for the tensile strength of soda-lime glass as a function of scratching load for dry air ignoring residual stress (heavy dashed line) and with residual stress (heavy solid line). For moist air the corresponding predictions are shown by the thin dashed and solid lines.

10.3 Surface Compressive Residual Stresses and Surface Flaw Detection

In the preceding discussion the residual stresses were associated with the formation of a subsurface flaw, and the fracture strength is greatly lowered by compressive residual stresses. Now we consider a surface flaw which exists before a compressive stress is introduced near the surface, for example by shot-peening. Figure 10.2 shows a typical distribution of the residual stress due to shot-peening. The high compressive residual stress near the surface greatly increases the resistance to fatigue crack initiation. However, when a surface flaw is already present, the compressive stress effectively closes the mouth of the crack and may prevent its detection by dye-penetrants, a process commonly used for detecting surface flaws. If the size of the flaw is larger than the depth of the zone of compressive stress, the crack tip will be under a tensile stress. To estimate the force required to open the surface flaw for detection, we need to compute the displacement caused by releasing the compressive stress near the surface.

Castigliano's theorem may be used to obtain the displacement due to a point load or the rotation due to a moment for a cracked body [123]. For two-dimensional parts of unit dimension perpendicular to the x-y plane subjected to mode I loading, a general expression for the displacements v on the surface at a distance s from the crack plane can be obtained by introducing a pair of virtual forces F at location $y = s$ as shown in Fig. 10.3. This leads to

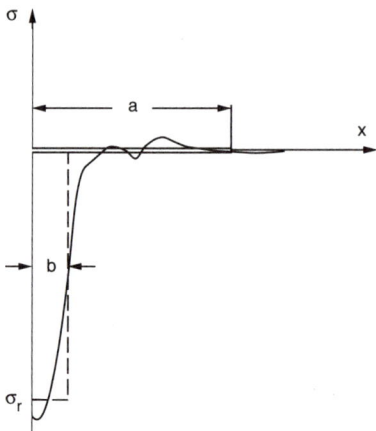

Fig. 10.2. Schematic of residual stress measured for a typical shot-peening application (solid line) and approximation by a rectangular distribution.

$$v(a, s) = \frac{\partial U}{\partial F}\Big|_{F=0} = \frac{1}{E'} \int_0^a K_I(a) \frac{\partial K_I^f(a, s)}{\partial F} da \qquad (10.2)$$

where a is the crack length, $E' = E$ for plane stress and $E/(1 - \nu^2)$ for plane strain with E and ν being the elastic modulus and Poisson's ratio respectively, U is the change of the strain energy due to the crack, K_I and K_I^f are the stress intensity factors for an arbitrary stress on the crack faces and the virtual force F respectively. Equation 10.2 is written in a general form which will be used later. For the present discussion, only the displacement at the mouth of the crack (i.e., half the crack opening) is of interest and $s = 0$. To simplify the computation we approximate the residual stress due to shot peening by a uniform compressive stress of magnitude σ_r, from $x = 0$ to b as shown in Fig. 10.3. The corresponding K_I is negative, and hence fictitious, but is needed to obtain the displacements required for subsequent calculations. The expression for K_I given in [17] is

$$K_I = 1.12\sigma_r(\pi a)^{\frac{1}{2}} f(\frac{b}{a}) \qquad (10.3)$$

with

$$f(\frac{b}{a}) = \begin{cases} 1 & a \leq b \\ 1 - (\frac{2}{\pi})(1 - \frac{3}{28}\frac{b}{a})\cos^{-1}(\frac{b}{a}) & a > b \end{cases}$$

The stress intensity factor K_I^f for a pair of crack closing line forces at the mouth of the crack, which is also negative, is given in [123] as

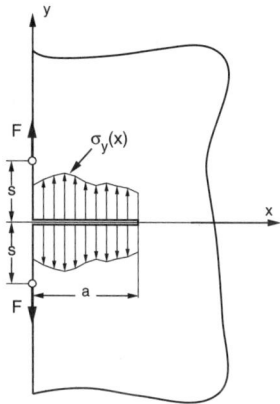

Fig. 10.3. A pair of virtual forces introduced at locations distance s from the crack plane to calculate the displacements due to the tractions on the crack faces.

$$K_I^f = 2.6\frac{F}{\sqrt{\pi a}} \tag{10.4}$$

and Eq. (10.2) becomes

$$v_r(a) = 2.9\frac{\sigma_r}{E'}aH(\frac{b}{a}) \tag{10.5}$$

where

$$H(\frac{b}{a}) = \frac{b}{a} + \int_{b/a}^{1}[1 - (\frac{2}{\pi})(1 - \frac{3}{28}\frac{b/a}{z})\cos^{-1}(\frac{b/a}{z})]dz, \quad a > b \tag{10.6}$$

In Eq. (10.6) the dummy variable z represents the ratio of any intermediate crack size to the final crack size. The displacement v_r would only appear when the crack is entirely opened by external loading.

To open the crack, we apply a uniform tensile stress so over the crack faces and the corresponding displacement v_o can be obtained by setting $b/a = 1$ in Eq. (10.5). Thus, the stress required for opening the crack is given by equating v_o and v_r to obtain

$$\sigma_o = \sigma_r H(\frac{b}{a}) \tag{10.7}$$

Figure 10.4 shows the ratio of σ_o/σ_r against b/a. It is seen that the opening stress required decreases as the size of the crack increases. Equation (10.7) is useful in showing the magnitude of the tensile stress required to open the crack so that it would be readily detectable.

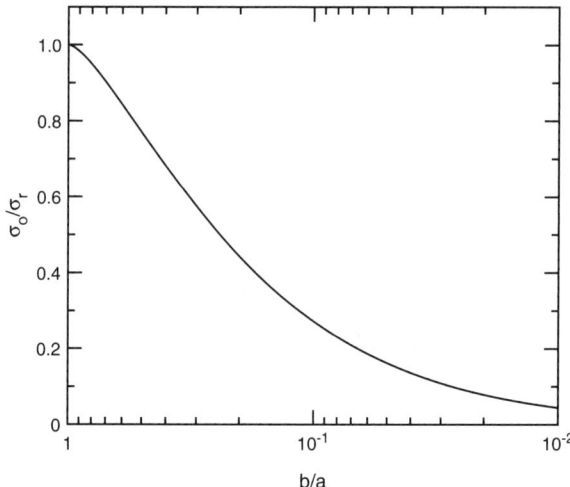

Fig. 10.4. The ratio of the opening stress σ_o to the residual stress σ_r as a function of b/a.

We now consider the more general problem in which the crack mouth is closed. If a is several times larger than b, it is reasonable to assume that the contact stress σ_q is uniform over the region $0 \leq x \leq b$. The corresponding crack mouth displacement is denoted by $v_q(a)$. The application of external loading alone, which leads to a uniform tensile stress σ_e over the entire crack, will produce a crack mouth displacement $v_e(u)$. A simplified approach, which neglects contact displacements in a solid body, leads to

$$v_q(a) = v_r(a) - v_e(a) \qquad (10.8)$$

The magnitude of the contact stress can be obtained from Eqs. (10.5) and (10.8) as

$$\sigma_q = \sigma_r - \frac{\sigma_e}{H(b/a)} \qquad (10.9)$$

Clearly, σ_q vanishes when Eq. (10.7) is satisfied, i.e. $\sigma_e = \sigma_o$. The stress intensity factor due to σ_q and σ_e is given by

$$K_I = 1.12\sqrt{\pi a}[\sigma_e + \sigma_q f(\frac{b}{a})]$$
$$= 1.12\sqrt{\pi a}\left\{\sigma_e + [\sigma_r - \frac{\sigma_e}{H(b/a)}]f(\frac{b}{a})\right\} \qquad (10.10)$$

which is valid only for $a > b$. Since surface treatments, such as shot peening, are often used in applications involving fatigue, it is useful to know the effect of residual stresses on the range of the stress intensity factor ΔK_I and on the mean K_I. Because the change of K_I is only caused by the change of applied stress we find from Eq. (10.10)

$$\frac{\Delta K_I}{\Delta K_{Io}} = 1 - \frac{f(b/a)}{H(b/a)} \quad with \quad \Delta K_{Io} = 1.12\Delta\sigma\sqrt{\pi a} \qquad (10.11)$$

which as shown in Fig. 10.5 starts at zero and approaches unity as b/a decreases. The mean value of K_I may be obtained from Eq. (10.10).

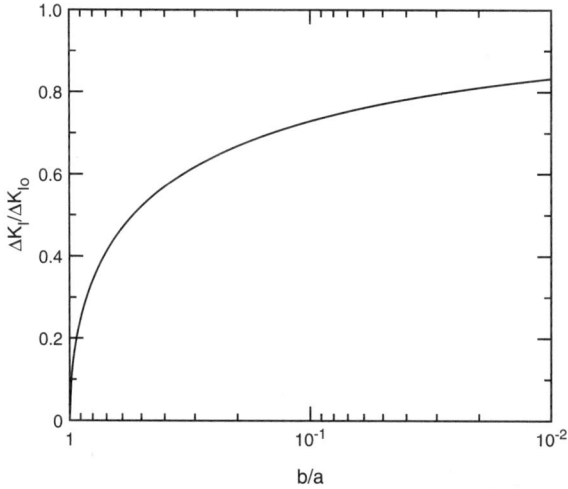

Fig. 10.5. The ratio of ΔK_I for a closed crack to the nominal ΔK_{Io} as a function of b/a.

Although the preceding discussion relates to edge cracks in a two-dimensional body, the computational procedure could be extended to other types of surface flaws.

The approach we have outlined for the calculation of K_I is simplified in several respects. It assumes that a pre-existing crack will not influence the residual stress due to shot peening. The contact stress is assumed to be confined to the region $0 \le x \le b$, and localized displacements due to contact stresses are ignored. However, we estimate that neglecting contact displacements will lead to errors of about 5% in estimating ΔK_I. A more detailed study of this problem would appear to be worthwhile.

10.4 Measurement of Stress Intensity Factors Using the Slitting Method

Traditionally, the measurement of stress intensity factors relies on obtaining the energy release rate estimated from the change of the area under load-displacement curves or the local deformation near the crack tip using techniques such as photoelastic coating. The first approach primarily applies to applications under external loads and the second one is time consuming and prone to the influence of steep strain gradient near the crack tip. In this section we will show that the crack compliance method can be conveniently used to estimate stress intensity factors as a function of a crack of increasing depth without the shortcomings of the other techniques[117].

To explain the basis of the crack compliance method we consider a strip, shown in Fig. 10.6, which is subjected to residual stress $\sigma_y(x)$. For near-surface measurement one or more strain gages are located close to the mouth of the crack. For through-thickness measurement a strain gage is located on the back face of the strip. In either case the normal strain $\epsilon(a, y)$ at a location $y = s$ produced by introducing a crack of depth a is given by differentiating Eq. (10.2) to obtain

$$\epsilon(a, s) = \frac{\partial v(a, s)}{\partial s} = \frac{\partial^2 U}{\partial F \partial s}_{|F=0} = \frac{1}{E'} \int_0^a K_I(a) \frac{\partial^2 K_I^f(a, s)}{\partial F \partial s} da \qquad (10.12)$$

in which $K_I(a)$ is the mode I stress intensity factor due to the residual stress and $K_I^f(s, a)$ is the stress intensity factor due to a virtual load F per unit of width at location s. Since strains can be measured very precisely with strain gages, Eq. (10.12) is more commonly used than one based on measurement of displacement.

Equation (10.12) only gives the strain at location s but in a test the strain is measured over a finite distance from s_1 to s_2. As shown in Appendix F for non-uniform elastic strain fields, the strain measured by a strain gage is essentially the average strain ϵ_m, and Eq.(10.12) in this case may be rewritten as

$$\epsilon_m(a) = \frac{1}{s_2 - s_1} \int_{s_1}^{s_2} \epsilon(a, s) ds$$

$$= \frac{1}{E'(s_2 - s_1)} \int_0^a K_I(a) \left[\int_{s_1}^{s_2} \frac{\partial^2 K_I^f(a, s)}{\partial F \partial s} ds \right] da \qquad (10.13)$$

Carrying out the integration over distance from s_1 to s_2, we find

$$\epsilon_m(a) = \frac{1}{E'(s_2 - s_1)} \int_0^a K_I(a) \left[\frac{\partial K_I^f(a, s_2)}{\partial F} - \frac{\partial K_I^f(a, s_1)}{\partial F} \right] da \qquad (10.14)$$

The availability of an analytical form of the crack compliance functions makes it possible to evaluate the stress intensity factor without having to obtain the residual stress. Taking a derivative of the average strain given in Eq. (10.14) with respect to the crack size a leads to

$$K_I(a) = E'(s_2 - s_1) \left[\frac{d\epsilon_m(a)}{da} \right] / \left[\frac{\partial K_I^f(a, s_2)}{\partial F} - \frac{\partial K_I^f(a, s_1)}{\partial F} \right] \qquad (10.15)$$

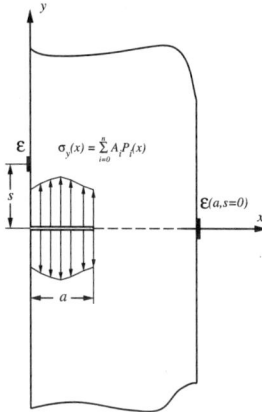

Fig. 10.6. A thin cut is introduced in a body to release stresses while strains are measured at selected locations.

Thus, the stress intensity factor due to residual stresses can be estimated directly from the change of strain measured when a thin cut is introduced. In general K_I^f can be expressed as

$$K_I^f(a, s) = \frac{F}{t} \sqrt{\frac{a}{\pi}} f^f(a, s) \qquad (10.16)$$

where t is the thickness. Equation (10.16) leads to

$$\frac{\partial K_I^f(a, s)}{\partial F} = \frac{1}{t} \sqrt{\frac{a}{\pi}} f^f(a, s) \qquad (10.17)$$

Equation (10.15) can now be rewritten as

$$K_I(a) = E't \sqrt{\frac{\pi}{a}} (s_2 - s_1) \left[\frac{d\epsilon_m(a)}{da} \right] / \left[f^f(a, s_2) - f^f(a, s_1) \right] \qquad (10.18)$$

Unlike the measurement of residual stresses, the measurement of stress intensity factors does not requires the use of a least squares fit. Its computation is more direct, which includes the differentiation of the measured strain variation and computation of $f^f(a, s_2) - f^f(a, s_1)$ at the location of the strain gage. For simple geometries such as a disk or a beam, the solution of $f^f(a, s)$ for a strain gage located on the back face directly opposite the cut have been obtained in Chapter 4. For a disk, if the gage length is very small compared with the disk diameter D, we have a very simple expression,

$$\frac{f^f(a/D, s_2) - f^f(a/D, s_1)}{s_2 - s_1} \approx \frac{4.48}{D(1 - a/D)^{3/2}} \tag{10.19}$$

which, when used in Eq. (10.15), gives

$$K_I(a) = \frac{E'D^2}{4.48}\sqrt{\frac{\pi}{a}}(1 - \frac{a}{D})^{3/2}\left[\frac{d\epsilon(a,0)}{da}\right] \tag{10.20}$$

For a beam of thickness t and a strain gage of length ℓ positioned δy from the centerline of the cut, as shown in Fig. 10.7 the solution for the average strain may be expressed in a general form as,

$$\frac{f^f(a/t, s_2) - f^f(a/t, s_1)}{s_2 - s_1} = \frac{f^f(a/t, \ell/2 + dy) - f^f(a/t, 0)}{\ell/2 + dy}$$

$$+ \frac{f^f(a/t, \ell/2 - dy) - f^f(a/t, 0)}{\ell/2 - dy}$$

$$= \frac{f^f(a/t, \ell/2 + dy)}{\ell/2 + dy} + \frac{f^f(a/t, \ell/2 - dy)}{\ell/2 - dy} \tag{10.21}$$

noting that $f^f(a/t, 0) = 0$ because it corresponds to the situation of a pair of point forces of equal magnitude but opposite direction acting at the same location on the plane of cut, and

$$\frac{f^f(a/t, \ell/2 \pm dy)}{\ell/2 \pm dy} = f_o(\frac{a}{t})\int_0^{a/t} f(x, \frac{a}{t})\frac{S(\ell/2 \pm dy, 1 - x)}{\ell/2 \pm dy}dx \tag{10.22}$$

Equation (10.18) for an edge-cracked beam now becomes

$$K_I(a) = \frac{E't}{f_o(a/t)}\sqrt{\frac{\pi}{a}}\frac{d\epsilon_m(a)}{da}$$

$$/ \int_0^{a/t} f(x, a/t)[\frac{S(\ell/2t + dy/t, 1 - x)}{\ell/2 + dy} + \frac{S(\ell/2t - dy/t, 1 - x)}{\ell/2 - dy}]dx \tag{10.23}$$

For simplicity, we take variables a, dy and ℓ as quantities normalized by thickness t and Eq.(10.23) may be simplified as

$$K_I(a) = \frac{E'}{f_o(a)}\sqrt{\frac{t\pi}{a}}\left[\frac{d\epsilon_m(a)}{da}\right]$$

$$/ \int_0^a f(x, a)\left[\frac{S(\ell/2 + dy, 1 - x)}{\ell/2 + dy} + \frac{S(\ell/2 - dy, 1 - x)}{\ell/2 - dy}\right]dx \tag{10.24}$$

in which $f_o(a)$ is given in Appendix A, $f(x, a)$ may be obtained from the solution of K_I also given in Appendix A, and $S(\ell/2 \pm dy, 1 - x)$ may be obtained from Appendix B. The integration in Eq.(10.24) requires a numerical computation and a C++ program is provided in Appendix G to facilitate the computation.

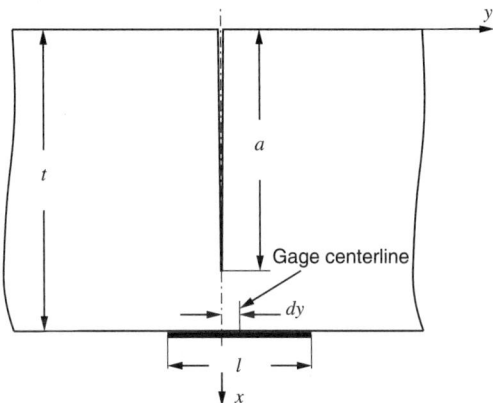

Fig. 10.7. Schematics of a strain gage positioned close to the centerline of the cut on the back face.

For other 2-D configurations, a solution will be usually unavailable and a numerical method such as FEM may be used to compute $f^f(a, s)$ in Eq. (10.16). As shown in [27], $f^f(a, s)$ for mode-I loading can be conveniently obtained from the displacement $v^f(a, s)$ at location s due to load F per unit of width at the same location, i.e.,

$$f^f(a, s) = \sqrt{\frac{\pi t^2}{a}} \sqrt{\frac{E'}{F} \frac{\partial v^f(a, s)}{\partial a}} = \sqrt{\frac{\pi t}{a}} \sqrt{\frac{\partial \hat{v}^f(a, s)}{\partial(a/t)}} \qquad (10.25)$$

in which $v^f(a, s)$ can be obtained using element meshes presented in Chapter 5 without utilizing any special elements near the crack tip and $\hat{v}^f$ is dimensionless and independent of the choice of E' and F. Substituting Eq. (10.25) into Eq. (10.18) for gage length ℓ and offset dy, we find

$$K_I(a) = E' \sqrt{t} \left[\frac{d\epsilon_m(a)}{da} \right]$$

$$\Big/ \left[\sqrt{\frac{\partial \hat{v}^f(a, \ell/2 + dy)}{\partial(a/t)}} \Big/ (\ell/2 + dy) + \sqrt{\frac{\partial \hat{v}^f(a, \ell/2 - dy)}{\partial(a/t)}} \Big/ (\ell/2 - dy) \right] (10.26)$$

Since the values of the $\partial \hat{v}^f(a, \ell/2 + dy)/\partial(a/t)$ are computed over a number of even-spaced depth of cuts, in practice an interpolation is usually required to determine the values at the actual depths of cut used for strain measurement.

10.5 Discussion

Griffith's classic work, which was presented in 1921, is revisited to show the influence of residual stresses on the tensile strength of glass.

The phenomenon of crack closure due to the residual stress in the wake of a propagating crack is familiar to those working in fatigue. However, the fact that surface compressive stresses may lead to closing of the crack mouth does require special attention in the area of non-destructive inspection for cracks.

The equivalency of energy release rate and the stress intensity factor, which is required to derive Eqs. (10.2) and (10.12), was shown in Irwin's classic paper in 1957. It was only until 1980's that the relation was used to obtain the deformation due to the release of residual stresses. In the last section we showed that the slitting method can also be used to measure stress intensity factors directly from the measured strain variations.

A

K_I and K_{II} Solutions for an Edge-Cracked Beam

A.1 An Expression for K_I

An expression of K_I for an edge-cracked beam is originally given in [17] as

$$K_I = \sqrt{\pi a t} f_o(a)\{\sigma(0) + \frac{2}{\pi}\int_0^a (1-z)^2 cos^{-1}[\frac{(1-a)z}{a(1-z)}][1 - \frac{z}{a}G(a)]\frac{\partial \sigma(z)}{\partial z}dz\}$$

(A.1)

in which [93]

$$f_o(a) = \left\{\frac{0.752 + 2.017a + 0.369[1 - \sin(\pi a/2)]^2}{\cos(\pi a/2)}\right\}\sqrt{\frac{2}{\pi a}}\tan(\frac{\pi a}{2}) \quad (A.2)$$

and

$$G(a) = \alpha(1 - 7a)(1 - a)^5$$

where $\alpha \approx 0.12/1.12 = 3/28$, a and z are the crack length and distance normalized by thickness. Equation (A.1) is only convenient when the derivative of the stress field can readily be evaluated. Computer programs written in C are provided in Appendix C to evaluate the derivative of Legendre and Chebyshev polynomials. For a general stress field we may rewrite Eq. (A.1) as

$$K_I = \sqrt{\pi a t}(\frac{2}{\pi})f_o(a)\int_0^a \left\{(1-z)(2 + G/a - 3zG/a)cos^{-1}[\frac{(1-a)z}{a(1-z)}]\right.$$

$$\left. + \frac{(1-a)(1-zG/a)}{a\sqrt{1 - \left[\frac{(1-a)z}{a(1-z)}\right]^2}}\right\}\sigma(z)dz \quad (A.3)$$

One problem with Eq. (A.3) is that it is not well defined for an edge crack in semi-infinite plane. It would be convenient to use a single solution to compute K_I for the whole range of the thickness. Letting $v = z/a$, the first term of the integrand of Eq. (A.3) becomes

$$\int_0^1 (1 - va)(2a + G - 3Gva) \cos^{-1}(\frac{(1-a)v}{1-av})\sigma(va)dv$$

and the second term becomes

$$\int_0^1 \frac{(1-a)(1-Gv)}{\sqrt{1 - [\frac{(1-a)v}{1-av}]^2}}\sigma(va)dv$$

Although evaluation of derivative is no longer needed, the second term contains a singularity at $v = 1$ that makes an accurate numerical integration difficult. To eliminate the singularity, we rewrite it as

$$\int_0^a \frac{(1-a)(1-Gv)}{\sqrt{1 - (\frac{(1-a)v}{1-av})^2}}\sigma(va)dv$$

$$= \int_0^{\pi/2} \frac{(1-a)^2[1 - a + (a - G)\sin u]}{(1 - a + a\sin u)^3}\sigma(\frac{a\sin u}{1 - a + a\sin u})du \qquad (A.4)$$

The final expression now takes the following form

$$K_I = \sqrt{\pi a t}(\frac{2}{\pi})f_o(a)\left\{ \int_0^1 (1 - va)(2a + G - 3Gva) \cos^{-1}[\frac{(1-a)v}{1-av}]\sigma(va)dv \right.$$

$$\left. + \int_0^{\pi/2} \frac{(1-a)^2[1 - a + (a - G)\sin u]}{(1 - a + a\sin u)^3}\sigma(\frac{a\sin u}{1 - a + a\sin u})du \right\} \qquad (A.5)$$

A.2 An Expression for K_{II}

The expression for K_{II} takes a similar form as that for K_I, i.e.,

$$K_{II} = \sqrt{\pi a t}g_o(a)\left\{[\tau(0)]\right.$$

$$\left. +(\frac{2}{\pi})\int_0^a (1 - z)\cos^{-1}(\frac{F(a)z}{1-z})[1 - \frac{z}{a}H(a)]\frac{\partial\tau(z)}{\partial z}dz \right\} \qquad (A.6)$$

where [123]

$$g_o(a) = (1.122 - 0.561a + 0.085a^2 + 0.18a^3)/(1 - a)^{1/2}$$

and

$$H(a) = \alpha(1 - 2.5a)(1 - a)$$

B

Stresses Due to Point Forces

The normal stress on the central plane of the strip due to a pair of horizontal or vertical point loads acting symmetrically at distance s on the upper edge of a strip is given respectively by

$$\frac{F}{\pi t}S(s,x) = \frac{F}{\pi t}\{\frac{4s^3}{(s^2+x^2)^2} + \int_0^\infty [(\alpha-1)G(\alpha,x) + \alpha H(\alpha,x)]e^{-\alpha}\sin(\alpha s)d\alpha\}$$

$$\frac{Q}{\pi t}T(s,x) = \frac{Q}{\pi t}\{\frac{4s^2 x}{(s^2+x^2)^2} + \int_0^\infty [\alpha G(\alpha,x) + (\alpha+1)H(\alpha,x)]e^{-\alpha}\cos(\alpha s)d\alpha\}$$

$$(B.1)$$

where

$$G(\alpha,x) =$$
$$\frac{\alpha(1-x)\sinh(\alpha x) - 2\cosh[(\alpha(1-x)] + \alpha x\sinh[\alpha(1-x)] - 2\cosh(\alpha x)}{\sinh(\alpha)+\alpha}$$
$$+\frac{\alpha(1-x)\sinh(\alpha x) + 2\cosh[\alpha(1-x)] - \alpha x\sinh[\alpha(1-x)] - 2\cosh(\alpha x)}{(\sinh(\alpha)-\alpha)}$$

$$H(\alpha,x) = \frac{\alpha(1-x)\cosh(\alpha x) - \sinh[\alpha(1-x)] + \alpha x\cosh[\alpha(1-x)] - \sinh(\alpha x)}{\sinh(\alpha)+\alpha}$$
$$+\frac{\alpha(1-x)\cosh(\alpha x) + \sinh[\alpha(1-x)] - \alpha x\cosh[\alpha(1-x)] - \sinh(\alpha x)}{\sinh(\alpha)-\alpha}$$

Replacing x by $1-x$ in Eq. (B.1) leads to expressions for the normal stresses due to point loads acting on the lower edge of the strip.

For strain computation the partial derivative with respect to s gives

$$\frac{\partial S(s,x)}{\partial s} = \frac{4s^2(3x^2-s^2)}{(s^2+x^2)^3} + \int_0^\infty [(\alpha-1)G(\alpha,x) + \alpha H(\alpha,x)]e^{-\alpha}\alpha\cos(\alpha s)d\alpha$$

$$\frac{\partial T(s,x)}{\partial s} = \frac{8sx(x^2-s^2)}{(s^2+x^2)^3} - \int_0^\infty [(\alpha-1)G(\alpha,x) + \alpha H(\alpha,x)]e^{-\alpha}\alpha\sin(\alpha s)d\alpha$$

$$(B.2)$$

for the horizontal and vertical point loads respectively.

For a pair of horizontal point loads on the lower edge

$$\frac{\partial S(1-x,s)}{\partial s} = \frac{4s^2[3(1-x)^2 - s^2]}{[s^2 + (1-x)^2]^3}$$

$$+ \int_0^\infty [(\alpha - 1)G(\alpha, 1 - x) + \alpha H(\alpha, 1 - x)]e^{-\alpha}\alpha \cos(\alpha s)d\alpha \qquad (B.3)$$

which, at $s = 0$, reduces to

$$\frac{\partial S(1-s,x)}{\partial s}\bigg|_{s=0} = \int_0^\infty [(\alpha - 1)G(\alpha, 1 - x) + \alpha H(\alpha, 1 - x)]e^{-\alpha}\alpha d\alpha \quad (B.4)$$

To simplify the computation of Eq. (B.4) we express it in terms of a Legendre polynomial expansion

$$\frac{\partial S(s, 1-x)}{\partial s}\bigg|_{s=0} = \frac{12}{\pi}L_1(x) + \sum_{i=2}^n b_i\, L_i(x) \qquad (B.5)$$

For $n = 15$ the coefficients b_i are tabulated in Table B.1.

Table B.1. Coefficients of Legendre series for approximation of Eq. (B.4)

i	b_i	i	b_i
2	-2.263850148374880e-001	9	2.430841835789998e 005
3	1.479050903091393e-001	10	-5.110357726474321e-006
4	-2.610313660458882e-002	11	1.106426725616064e-006
5	9.050406633413190e-003	12	-2.283685583449879e-007
6	-1.860168650373919e-003	13	4.738479609037172e-008
7	4.934447220567688e-004	14	-9.591872844605387e-009
8	-1.045400500176257e-004	15	1.935575313092089e-009

C

C Subroutines for the Calculation of Polynomial Series

C.1 Chebyshev Polynomials

/* This program uses the recurrence relation to compute the value of an nth order Chebyshev polynomial series for a given value of x.

    Input variables:

        x - independent variable with $0 \leq x \leq 1$;

        n - order of the series;

        a[] - coefficients vector.*/

```
double ChbySum(double x,int n,double a[])
{
    double v0,v1,v2,sum;
    int j;
    sum = a[0];
    x = 2.*x-1.;
    if(n>0)
    {
        v0 = 1.;
        v1 = x;
        sum += a[1]*v1;
        for(j=2; j<=n; ++j)
        {
            v2 = 2*x*v1-v0;
            sum = sum+a[j]*v2;
            v0 = v1;
            v1 = v2;
        }
        return sum;
    }
    return sum;
}
```

/* This program uses the recurrence relation to compute the value of an nth order Chebyshev polynomial for a given value of x.
   Input variables:
        x - independent variable with $0 \le x \le 1$;
        n - order of the polynomial;*/

```c
double ChbyVal(double x,int n)
{
    double v0,v1,v2;
    int j;
    if(n>0)
    {
        x = 2.*x-1.;
        v0 = 1.;
        v1 = x;
        for(j=2; j≤n; ++j)
        {
            v2 = 2*x*v1-v0;
            v0 = v1;
            v1 = v2;
        }
        return v1;
    }
    return 1.;
}
```

/* This program uses the recurrence relation to compute the value of the first derivative of an nth order Chebyshev polynomial for a given value of x, which can be used directly in Eq. (A.1).
   Input variables:
        x - independent variable with $0 \le x \le 1$;
        n - order of the polynomial;
This routine can be efficiently used in the numerical integration of the LEFM solution.*/

```c
double ChbyDVal(double x,int n)
{
    double v0,v1,v2;
    int j;
    x = 2.*x-1.;
    if(n>0)
    {
        n--;
        if(n>0)
```

```
    {
        v0 = 1.;
        v1 = 2*x;
        for(j=2; j≤n; ++j)
        {
            v2 = 2*x*v1-v0;
            v0 = v1;
            v1 = v2;
        }
        return v1*(n+1);
    }
    return 1.;
}
return 0.;
}
```

C.2 Legendre Polynomials

```
/* This program uses the recurrence relation to compute the value of an nth
order Legendre series for a given value of x.
    Input variables:
        x - independent variable with 0 ≤ x ≤ 1;
        n - order of the series;
        a[] - coefficients vector.
A continuous normal residual stress distribution can be obtained by setting
a[0] and/or a[1] to zero.*/
double LegenSum(double x,int n,double a[])
{
    double v0,v1,v2,sum;
    int j;
    sum = a[0];
    x = 2.*x-1.;
    if(n>0)
    {
        v0 = 1.;
        v1 = x;
        sum += a[1]*v1;
        for(j=2; j≤n; ++j)
        {
            v2 = (2*j-1.)*x*v1-(j-1.)*v0/j;
            sum = sum+a[j]*v2;
            v0 = v1;
            v1 = v2;
        }
        return sum;
    }
    return sum;
}


/* This program uses the recurrence relation to compute the value of an
nth order Legendre polynomial for a given value of x.
    Input variables:
        x - independent variable with 0 ≤ x ≤ 1;
        n - order of the polynomial;*/
double LegenVal(double x,int n)
{
    double v0,v1,v2;
    int j;
    if(n>0)
    {
        x = 2.*x-1.;
```

```
        v0 = 1.;
        v1 = x;
        for(j=2; j≤n; ++j)
        {
            v2 = (2*j-1.)*x*v1-(j-1.)*v0/j;
            v0 = v1;
            v1 = v2;
        }
        return v1;
    }
    return 1.;
}
```

/* This program returns the value of the first derivative of an nth order Legendre polynomial for a given value of x, which can be used directly in Eq. (A.1).

Input variables:

x - independent variable with $0 \leq x \leq 1$;

n - order of the polynomial;

This routine can be efficiently used in the numerical integration of the LEFM solution.*/

```
double LegenDVal(double x,int n)
{
    double v0,v1,v2,d0,d1;
    int j;
    x = 2.*x-1.;
    if(n>0)
    {
        v0 = 1.;
        v1 = x;
        d0 = 1.;
        d1 = 3.*x;
        for(j=2; j≤n; ++j)
        {
            v2 = (2*j-1.)*x*v1-(j-1.)*v0/j;
            d1 = j*v1+x*d0;
            v0 = v1;
            v1 = v2;
            d0 = d1;
        }
        return d0;
    }
    return 0.;
}
```

C.3 Jacobi Polynomials

A Jacobi polynomial series can be used to represent a continuous shear stress which vanishes at free surfaces ($x = 0$ and $x = 1$), i.e.,

$$\tau(x) = x(1 - x) \sum_{i=1}^{n} A_i J_i(x)$$

A residual shear stress is obtained when $A_1 = 0$

```
/* This program uses the recurrence relation to compute the value of an
nth order Jacobi polynomial for a given value of x.
    Input variables:
        x - independent variable with 0 ≤ x ≤ 1;
        n - order of the polynomial;
*/
double JacoVal(double x,int n)
{
    double v0,v1,v2;
    int j;
    x = 2.*x-1.;
    if(n>0)
    {
        v0 = 1.;
        v1 = 2.*x;
        for(j=2; j≤n; ++j)
        {
            v2 = ((2*j+1)*(j+1)*x*v1-j*(j+1)*v0)/(j*(2+j));
            v0 = v1;
            v1 = v2;
        }
        return v1;
    }
    return 1.;
}

/* This program returns the sum of an nth order Jacobi polynomial series
for a given value of x.
    Input variables:
        x - independent variable with 0 ≤ x ≤ 1;
        n - order of the polynomial;
*/
double JacoSum(double x,int n,double A[])
{
    double v0,v1,v2,sum;
```

```
    int j;
    x = 2.*x-1.;
    if(n>0)
    {
        v0 = 1.;
        v1 = 2.*x;
        for(j=2; j≤n; ++j)
        {
            v2 = ((2*j+1)*(j+1)*x*v1-j*(j+1)*v0)/(j*(2+j));
            v0 = v1;
            v1 = v2;
        }
        return v1;
    }
    return 1.;
}
```

D

K_I Solution for an Edge-Cracked Disk

The K_I solution for an edge-cracked disk due to an arbitrary loading on crack faces has been studied by several authors using weight functions [132, 49]. As pointed out by Schindler [118], for a deep crack subjected to self-equilibrating stresses, the polynomial approximation proposed by Petroski and Achenbach [92] is no longer applicable because of the inconsistency between the far-field conditions and the near crack tip functions used in the polynomial. To overcome this problem, a new weight function is recently proposed by Schindler [118] which leads to an improved result for deep cracks. Here we take an alternative approach to obtain a weight function based on an asymptotic interpolation of the solutions for a very shallow and a very deep crack.

D.1 Analysis

Consider a disk of diameter D with an edge crack of size a normalized by the diameter shown in Fig. D.1. For a very shallow crack the K_I solution reduces to that of for an edge crack in a semi-infinite plane. For an arbitrary normal stress $\sigma(x)$ on the crack faces, the solution has been given in [17] as

$$K_I = 1.12\sigma\sqrt{\pi a D} \int_0^a \frac{\partial}{\partial x} \left[-(\frac{2}{\pi}) \cos^{-1}(\frac{x}{a})(1 - \frac{\alpha x}{a}) \right] \sigma(x) dx \qquad (D.1)$$

in which x is the distance taken from the crack mouth, as shown in Fig. D.2, and α a constant approximately equal to $3/28$.

Following the same consideration for an edge-cracked strip [17], the solution for an arbitrary normal stress acting on the faces of a very deep crack in a disk may be given by

$$K_I = \sqrt{\pi a D} f_0(a) 2 \int_0^1 (1 - z)\sigma(z) dz \qquad (D.2)$$

where z is the normalized distance x/D. The K_I solution $f^0(a)$ for a uniform normal traction on crack faces shown in Fig. D.1 is given in a very simple

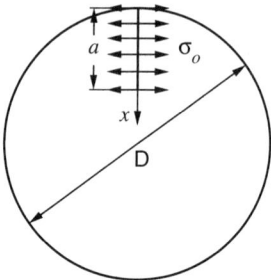

Fig. D.1. A disk of diameter D with an edge crack of size a subjected to a uniform stress σ_o on crack faces.

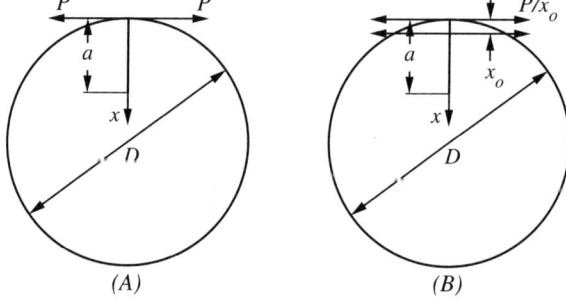

Fig. D.2. A disk of diameter $D = 1$ with an edge crack of size a subjected to a pair of forces P on the crack mouth (a), which is approximated by a strip load of P/x_o in (b).

form [49] as

$$f_0(a) = 1.12(1 - a)^{-3/2} \tag{D.3}$$

To obtain a solution of K_I for all values of a, the corresponding expression based on an asymptotic interpolation may be written as

$$K_I = \sigma\sqrt{\pi a D} f_0(a) f(a) \tag{D.4}$$

where σ is a reference stress and

$$f(a) = \int_0^a \frac{\partial}{\partial z}\left\{(1-(1-z)^2(\frac{2}{\pi})\cos^{-1}\left[\frac{z(1-a)}{(1-z)a}\right]H(z,a)\right\}\frac{\sigma(z)}{\sigma}dz \quad (D.5)$$

Eq. (D.4) reduces to Eq. (D.1) or (D.2) only if the asymptotic interpolation function $H(z,a)$ satisfies

$$\lim_{d\to\inf} H(z,a) = 1 - \alpha\frac{x}{a}\lim_{a\to d} H(z,a) = 1 \quad (D.6)$$

As will be shown later, the simplest form for $H(z,a)$ may be given by

$$H(z,a) = 1 - \alpha z(1-a)/a \quad (D.7)$$

To show that Eq. (D.7) leads to an excellent approximation of the exact K_I solution, we consider the solution for a pair of point forces acting at the edge of the crack as shown in Fig. D.2-a. For simplicity without losing generality, we take $D = 1$ in the derivation that follows. The exact solution of K_I in this case is given in [132] as

$$K_I = \frac{P}{\sqrt{\pi a}}\frac{2.5935 + 4.4533a}{(1-a)^{3/2}} \quad (D.8)$$

Replacing the stress field in Eq. (D.5) by a strip load P/x_o shown in Fig. D.2-b, Eq. (D.4), when combine with Eq. (D.7), becomes

$$K_I = P\sqrt{\pi a}\frac{f_0(a)}{x_o}\left\{1-(1-x_o)^2(\frac{2}{\pi})\cos^{-1}\left[\frac{x_o(1-a)}{a(1-x_o)}\right][1-\frac{\alpha x_o}{a}(1-a)]\right\} \quad (D.9)$$

Taking the limit of x_o approaching zero gives

$$K_I = \frac{1.12P}{(1-a)^{3/2}}\sqrt{\frac{\pi}{a}}[\alpha + \frac{2}{\pi} + a(2-\alpha-\frac{2}{\pi})] = \frac{P}{\sqrt{\pi a}}\frac{2.6170 + 4.4202a}{(1-a)^{3/2}} \quad (D.10)$$

The maximum difference between Eq. (D.10) and Eq. (D.8) is only 0.9%. Turning to Eq. (D.7), we could also express the asymptotic interpolation function in a more general form. That is, for $D = 1$,

$$H(z,a) = 1 - \alpha z(1-a)(1+\sum_{i=1}^n c_i a^i)/a \quad (D.11)$$

where c_i are coefficients to be determined. To obtain the solution for the same problem shown in Fig. 2-b, we substitute Eq. (D.11) into Eq. (D.5) and replace the stress field by P/x_o to obtain

$$K_I = P\sqrt{\pi a}\frac{f_0(a)}{x_o}\left\{1-(1-x_o)^2(\frac{2}{\pi})\cos^{-1}\left[\frac{x_o(1-a)}{a(1-x_o)}\right]\right.$$
$$\left.[1-\alpha\frac{x_o}{a}(1-a)(1+\sum_{i=1}^n c_i a^i)]\right\} \quad (D.12)$$

Taking the limit of x approaching zero leads to

$$K_I = \frac{1.12P}{(1-a)^{3/2}} \sqrt{\frac{\pi}{a}} [\alpha + \frac{2}{\pi} + a(2 - \alpha - \frac{2}{\pi}) + \alpha(1-a) \sum_{i=1}^{n} c_i a^i] \quad (D.13)$$

A comparison of Eq. (D.13) with Eq. (D.8) indicates that all higher order terms of a_i with $i > 1$ in Eq. (D.11) are redundant and coefficients c_i must vanish. Substituting Eqs. (D.3) and (D.7) into Eq. (D.4), the K_I solution may be expressed in a very compact form as

$$K_I = \frac{1.12\sqrt{\pi a}}{(1-a)^{3/2}} \int_0^a \frac{\partial}{\partial z} \left\{ (1 - (1-z)^2 (\frac{2}{\pi}) \cos^{-1} \right.$$
$$\left. \left[\frac{z(1-a)}{(1-z)a} \right] [1 - \alpha z(1-a)/a] \right\} \sigma(z) dz \quad (D.14)$$

D.2 Results

Two loading conditions are considered. The first one is a parabolic stress distribution given by

$$\sigma(z) = \sigma_2 (1 - 2z)^2 \quad (D.15)$$

The corresponding values of K_I have been obtained by Gregory [59] using an analytical solution. The second one is a linear stress distribution given by

$$\sigma(z) = \sigma_1(1 - 2z) \quad (D.16)$$

The values of K_I can also be obtained by using the weight functions given by Schindler [118] or by Fett and Munz [49].

Substituting Eqs. (D.15) and (D.16) into Eq. (D.14), the values of K_I/σ are computed and tabulated in Tables D.1 and D.2 respectively. The results obtained by Gregory and given by Schindler or Fett and Munz using weight function are also tabulated in the two tables. It is seen that the agreement with the analytical solution by Gregory is excellent and the agreement with the results by Schindler or Fett and Munz is also very good.

Table D.1. Comparison of Eq. (D.14) and exact solution [59] for a parabolic stress given by Eq. (D.15)

a/d	Eq.(D.14)	Ref.[59]	Eq. (14)/Ref. [59]
0.05	0.424592	0.425181	0.998614
0.10	0.577839	0.578924	0.998126
0.15	0.686358	0.688234	0.997274
0.20	0.776176	0.779163	0.996166
0.25	0.860120	0.864490	0.994944
0.30	0.947400	0.953339	0.993771
0.35	1.046424	1.053997	0.992815
0.40	1.166271	1.175407	0.992228
0.45	1.318009	1.328485	0.992115
0.50	1.516399	1.527838	0.992513
0.55	1.782608	1.794490	0.993379
0.60	2.149051	2.160730	0.994595
0.65	2.668774	2.679536	0.995983
0.70	3.435373	3.444533	0.997341
0.75	4.629964	4.637054	0.998471
0.80	6.648814	6.653977	0.999224
0.85	10.52746	10.53239	0.999532
0.90	19.88281	19.89363	0.999456
0.95	57.74163	57.78478	0.999253

Table D.2. Comparison of Eq. (D.14) with solutions [49] and [118] for a linear stress given by Eq. (D.16)

a/d	Eq.(D.14)	Ref. [118]	Ref.[49]
0.05	0.450909	0.45065	0.45078
0.10	0.650149	0.65084	0.65065
0.15	0.815304	0.81710	0.81646
0.20	0.968727	0.97218	0.97086
0.25	1.120928	1.12682	1.12443
0.30	1.279455	1.28879	1.28478
0.35	1.451465	1.46547	1.45905
0.40	1.645100	1.66523	1.65540
0.45	1.870766	1.89873	1.88417
0.50	2.142858	2.18062	2.15970
0.55	2.482530	2.53239	2.50303
0.60	2.922670	2.98730	2.94682
0.65	3.517533	3.60014	3.54488
0.70	4.363147	4.46769	4.39225
0.75	5.645298	5.77700	5.67239
0.80	7.769517	7.93599	7.78448
0.85	11.79163	12.0058	11.7643
0.90	21.38246	21.6732	21.1986
0.95	59.78601	60.2655	58.6965

E

Stress Variation With the Location of the Virtual Forces on a Disk

Consider a pair of opposite forces F acting along a vertical chord $\overline{AB}$ of a disk shown in Fig. E.1. Assuming that each of the forces produces a simple radial stress distribution[124], the normal stress σ'_y at location C along a horizontal line $\overline{MN}$ passing the center of the disk O is given by

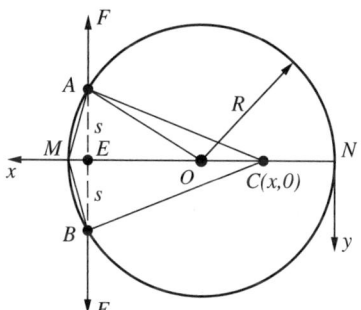

Fig. E.1. A disk of diameter $D = 2R$ subjected to a pair of forces F along chord $\overline{AB}$ of length $2s$.

$$\sigma'_y(x,0) = \frac{4F}{\pi s}(1 - \cos^2 \beta) \cos^2 \beta \qquad \text{(E.1)}$$

where s is the distance from F to $\overline{MN}$ and β is the angle between chord $\overline{AB}$ and the radius from F to location C. In rectangular coordinates with origin located at N, $\cos^2 \beta$ in Eq. (E.1) is given as

$$\cos^2 \beta = \frac{s^2}{R^2 + (R-s)^2 + 2(R-x)\sqrt{R^2 - s^2}} \tag{E.2}$$

where R is the radius of the disk and x is the distance from C to N. It is shown in [124] that Eq. (E.1) produces a uniform tension around the disk. For a stress-free boundary we need to superpose a uniform tension around the disk with an opposite sign, which is given by

$$\sigma = \frac{2F}{\pi D} \sin(\theta + \theta_1) \tag{E.3}$$

where θ and θ_1 are angles between chord $\overline{AB}$ and the distance from A or B to a location on the rim of the disk. Noticing $\theta + \theta_1$ is constant, we can write the corresponding normal stress on plane $\overline{MN}$ as

$$\sigma_y''(x, 0) = \frac{4F}{\pi D} \sin \theta \cos \theta = \frac{4F}{\pi D} \frac{\overline{ME}\ \overline{AE}}{\overline{AM}^2} = \frac{4F}{\pi D} \frac{s}{2R} \tag{E.4}$$

where θ is the angle between $\overline{AB}$ and $\overline{AM}$ or $\overline{BM}$. Carrying out the partial differentiations with respect to F and s, and setting s to zero leads to[1].

$$\frac{\partial^2 \sigma_y(x, 0)}{\partial F \partial s}\bigg|_{s=0} = \frac{\partial^2 \sigma_y'(x, 0)}{\partial F \partial s}\bigg|_{s=0} + \frac{\partial^2 \sigma_y''(x, 0)}{\partial F \partial s}\bigg|_{s=0} = 0 + \frac{4P}{\pi D^2} \tag{E.5}$$

---

[1] The result of Eq. (E.5) was first obtained during the work of [39]

F

Nonuniform Strain Measured Over a Finite Length of an Electric Resistance Strain Gage

Electric resistance strain gages have been used in many applications because of their accuracy and convenience. However, in situations where the length of a strain gage is large compared with the size of a region with a rapidly varying strain gradient, a question arises whether the relation based on a uniform strain field is still valid. In this note we will show that the relation is still valid provided that the average strain over the length of the gage is used.

When a uniaxial load is applied to a conductor of length L_o and cross-section area $A_o = w_o t_o$, the change of the resistance may be related to the strain ϵ, Poisson's ratio ν and specific resistance p by [41]

$$\frac{dR}{R_o} = \epsilon(1 + 2\nu) + \frac{dp}{p} \tag{F.1}$$

assuming ϵ is small and uniform.

In what follows we will demonstrate that, if the strain varies along the length, Eq. (F.1) is still valid provided that ϵ is replaced by the average strain ϵ_m. Since the specific resistance is much less influenced by the dimensional change of the strain gage, we replace p by p_m, which is uniform along the length and only dependent on ϵ_m.

The expression for resistance is now given by

$$R = p_m \int_0^{L_0 + \Delta L} \frac{ds}{A(s)} + dp \frac{L_0}{A_0} \tag{F.2}$$

in which s is the distance in the longitudinal direction. Assuming a rectangular cross section of width w and thickness t, the area A is given by

$$A = w \times t = A_0(1 + \frac{dw}{w_0})(1 + \frac{dt}{t_0}) = A_0[1 - 2\nu\epsilon(s) + \nu^2\epsilon^2(s)] \tag{F.3}$$

The change of resistance is thus given by

$$\frac{R - R_0}{R_0} = \int_0^{L_0 + \Delta L} \frac{(ds/L_0)}{1 - 2\nu\epsilon(s) + \nu^2\epsilon^2(s)} - 1 + \frac{dp}{p_m}$$

$$= \int_0^{1+\epsilon_m} \frac{ds'}{1 - 2\nu\epsilon(s') + \nu^2\epsilon^2(s')} - 1 + \frac{dp}{p_m} \tag{F.4}$$

in which $s' = s/L_o$. To check Eq. (F.4), we let ϵ be a constant, i.e.,

$$\int_0^{1+\epsilon} \frac{ds'}{1 - 2\nu\epsilon + \nu^2\epsilon^2} - 1 = \frac{1+\epsilon}{1 - 2\nu\epsilon + \nu^2\epsilon^2} - 1$$

$$= \frac{\epsilon + 2\nu\epsilon - \nu^2\epsilon^2}{1 - 2\nu\epsilon + \nu^2\epsilon^2} \approx \epsilon(1 + 2\nu) \tag{F.5}$$

which, when substituted into Eq. (F.4), leads to an expression identical to Eq. (F.1).

Consider an nth order nonuniform strain field which may take a form as

$$\epsilon_n(s') = \epsilon_m(n+1)(s')^n \tag{F.6}$$

where ϵ_m is the average strain of $\epsilon_n(s')$. Using Eq. (F.6), the first term in Eq. (F.4) becomes

$$\int_0^{1+\epsilon_m} \frac{ds'}{1 - 2\nu\epsilon_m(1+n)(s')^n + \nu^2\epsilon_m^2(1+n)^2(s')^{2n}} - 1 \tag{F.7}$$

As a comparison we tabulate the values of the ratio of Eq. (F.7) to the first term of Eq. (F.1) in Table F.1 for various values of ν, ϵ_m and n. It is seen that the difference varies slightly with different values of n and ν, and it is nearly proportional to ϵ_m. It should be pointed out that if any uniform strain is included in Eq. (F.6) the combined differences will be smaller than those tabulated in Table F.1. Thus, we may conclude that for elastic deformation Eq. (F.1) is sufficiently accurate to obtain the average strain over a region of nonuniform strain. In other words, the strain measured by an electrical-resistance strain gage represents closely the average strain over the length of the gage.

Table F.1. Ratio of Eq. (F.7) to Eq. (F.1) for different values of ϵ_m and ν for $n = 1$ to $n = 3$

$n = 1$	$\epsilon_m = 0.1$	$\epsilon_m = 0.01$	$\epsilon_m = 0.001$	$\epsilon_m = 0.0001$
$\nu = 0.25$	1.09347	1.00843	1.00083	1.00008
$\nu = 0.30$	1.11081	1.00987	1.00098	1.00010
$\nu = 0.35$	1.12084	1.01127	1.00111	1.00011
$n = 2$	$\epsilon_m = 0.1$	$\epsilon_m = 0.01$	$\epsilon_m = 0.001$	$\epsilon_m = 0.0001$
$\nu = 0.25$	1.15005	1.01248	1.00123	1.00012
$\nu = 0.30$	1.17872	1.01458	1.00143	1.00014
$\nu = 0.35$	1.20761	1.01661	1.00163	1.00016
$n = 3$	$\epsilon_m = 0.1$	$\epsilon_m = 0.01$	$\epsilon_m = 0.001$	$\epsilon_m = 0.0001$
$\nu = 0.25$	1.21831	1.01663	1.00161	1.00016
$\nu = 0.30$	1.26245	1.01941	1.00189	1.00019
$\nu = 0.35$	1.30780	1.02209	1.00215	1.00021

C++ Programs for the Calculation of Eq. (10.24)

Following the derivation of Eq. (A.5) we have

$$\int_0^a f(x,a) \left[\frac{S(\Delta s, 1-x)}{\Delta s} \right] dx$$

$$= \int_0^1 (1-va)(2a + G - 3Gva) \cos^{-1}[\frac{(1-a)v}{1-av}] \frac{S(\Delta s, 1-va)}{\Delta s} dv$$

$$+ \int_0^{\pi/2} \frac{(1-a)^2[1-a+(a-G)\sin u]}{(1-a+a\sin u)^3} \frac{S(\Delta s, (1-a)/(1-a+a\sin u))}{\Delta s} du \quad (G.1)$$

The numerical integration of Eq. (G.1) is encapsulated in a C++ class, PointF. A sample C++ program is provided to show the usage of PointF. The program has been tested using the Microsoft® Visual studio.net C++ 2005.

G.1 PointF Class Header – Listing of PointF.h

```
#define GSQN 30 // number of Gaussian quadrature points
class PointF
{
public:
    PointF(void);
    ~PointF(void);

// Gaussian-Lagrrier quadrature routine
    double infintg(double (PointF::*grand)(double));
// integrand for Gaussian-Lagrrier quadrature
    double integrandS(double x);
// Set gage length gl and offset dy for computing ave. strain
    void SetGPos(double x,double y) {gl=x; dy = y;};
    void SetNVal(int i);
```

```
// Compute stress due to F
   double FunS(double y);
////////////////////////////////////////////////////////////
// A modified Gaussian quadrature
   double intek(double a, double b,double (PointF::*grand)(double));
   double Fun1(double v);
   double Fun2(double u);
   double FunF(double a);
private:
   int n; // number of Gaussian-Lagrrier quadrature points
   static double ab[GSQN];
   static double wb[GSQN];
   static double add[GSQN];
   static double wkb[GSQN];
   double (PointF::*f)(double x);
   double ao; // current normalized depth of cut
   double Ga; // current value of G(a)
   double yo; // normalized distance on cut plane for stress due to F
   double so; // distance from the position of F to the plane of cut
   double gl; // Gage length
   double dy; // Offset of gage center
};
```

G.2 Code for Class Definition – Listing of PointF.cpp

```
#include ''pointf.h"
#define _USE_MATH_DEFINES
#include <math.h>

PointF::PointF(void)
:n(21), // accurate to at least 1.0e-9 for n = 21
ao(0), Ga(0) { }

PointF::~PointF(void) { }

////////////////////////////////////////////////////////////
double PointF::ab[GSQN] =
{0.234526109519618537, 0.576884629301886424, 1.07244875381781771,
1.72240877644464545, 2.52833670642579507, 3.49221327302199439,
4.61645676974973095, 5.90395850417717807, 7.35812673158943653,
8.98294092506170375, 10.7830186325538887, 12.7636979795992374,
14.9311397434474381, 17.2924543449313719, 19.8558608677867421,
22.6308889326376814, 25.6286360941471892, 28.8621018735170335,
32.3466291442543635, 36.1004948095275791, 40.1457197693999658,
44.5092079943361867, 49.2243949878495185, 54.3337213335559171,
59.8925091630914400, 65.9753772877550561, 72.6876280911816123,
```

```
80.1874469777939813, 88.735340417914390, 98.8295428682851138};
double PointF::wb[GSQN] =
{0.21044310793881323, 0.23521322966984802, 0.19590333597288113,
0.12998378628607033, 0.070578623865717035, 0.031760912509175152,
0.011918214834847041, 0.0037388162946272654, 9.8080331147953040e-4,
2.1486491849068858e-4, 3.9203419695654058e-5, 5.9345416585029408e-6,
7.4164047196725324e-7, 7.6045677945004026e-8, 6.3506043286326967e-9,
4.2813811668291118e-10, 2.3058991426852954e-11, 9.7993789867899799e-13,
3.2378020233599416e-14, 8.1718229928234848e-16, 1.5421338096502519e-17,
2.1197923331866732e-19, 2.0544296755444497e-21, 1.34698259129703856e-23,
5.6612941065144460e-26, 1.4185605452783491e-28, 1.9133754915490041e-31,
1.1922487604782053e-34, 2.6715112191015198e-38, 1.3386169421024840e-42};
/* A Gaussian-Laguerre quadrature based on 32 nodes. Only the first
n nodes and corresponding weights are used in the program. User
can change the number of nodes used in the program by specifying
the value of computation n. */
double PointF::infintg(double (PointF::*grand)(double))
{
    int i;
    double vk,*pa,*pw;

    pa = ab;
    pw = wb;
    vk = (this-->*grand)(0.044489365833267)*0.109218341952385;
    for(i=1;i<=n;++i)
        vk += (this-->*grand)(*pa++)**pw++;
    return vk;
}
void PointF::SetNVal(int i)
{
    if(i>30)
        n = 30;
    else
        n = i;
}

double PointF::FunS(double y)
{
    double t,x,S1,S2;
    yo = y;
    x = 1.0-yo;
    so = gl/2.0+dy; // distance from centerline
    t = so*so+x*x;
    f = &PointF::integrandS;
    S1 = infintg(f)/so+4.0*so*so/(t*t);

    so = gl/2.0-dy; // distance from centerline
    t = so*so+x*x;
    f = &PointF::integrandS;
```

```
    S2 = infintg(f)/so+4.0*so*so/(t*t);
    return S1+S2;
}
// The integrand for stress due to horizontal point forces on lower edge
double PointF::integrandS(double x)
{
    double sh,chy,shy,xy,xy1,shy1,chy1,t1,t2,t3,t4;
    xy = x*(1.0-yo);
    sh = sinh(x);
    chy = cosh(xy);
    shy = sinh(xy);
    xy1 = x*yo;
    shy1 = sinh(xy1);
    chy1 = cosh(xy1);
// G(a,1-x)
    t1 = (xy1*shy-2*chy1+xy*shy1-2*chy)/(sh+x);
    t2 = (xy1*shy+2*chy1-xy*shy1-2*chy)/(sh-x);
    t1 = (x-1)*(t1+t2);
// H(a,1-x)
    t3 = (xy1*chy-shy1+xy*chy1-shy)/(sh+x);
    t4 = (xy1*chy+shy1-xy*chy1-shy)/(sh-x);
    t3 = x*(t3+t4);
    return (t1+t3)*sin(so*x);
}
///////////////////////////////////////////////////////////
double PointF::add[GSQN] =
{3.8220118431826414e-4, 2.0127000924398829e-3, 4.9416273837414745e-3,
9.1619943579814602e-3, 1.4662870583408546e-2, 2.1429924043507954e-2,
2.9445506659319426e-2, 3.8688709308522369e-2, 4.9135418762999415e-2,
6.0758381392559484e-2, 7.3527274576168277e-2, 8.7408785945670025e-2,
1.0236670035882018e-1, 1.1836199441384390e-1, 1.3535293827526745e-1,
1.5329520455275442e-1, 1.7214198395246456e-1, 1.9184410740103914e-1,
2.1235017432458469e-1, 2.3360668674853737e-1, 2.5555818886887394e-1,
2.7814741173073420e-1, 3.0131542263712170e-1, 3.2500177889796658e-1,
3.4914468551848464e-1, 3.7368115641547325e-1, 3.9854717874097075e-1,
4.2367787988459235e-1, 4.4900769671886297e-1, 4.7447054664601283e-1};
double PointF::wkb[GSQN] =
{9.8072668083514134e-4, 2.2804620030062086e-3,3.5761774958745448e-3,
4.8627309151780669e-3, 6.1366317539060523e-3, 7.3945329424689573e-3,
8.6331464938068718e-3, 9.8492388730505907e-3, 1.1039636574159522e-2,
1.2201233593772101e-2, 1.3330999262075445e-2, 1.4425986044091701e-2,
1.5483337184198697e-2, 1.6500294137953705e-2, 1.7474203758266676e-2,
1.8402525211577409e-2, 1.9282836603504086e-2, 2.0112841295499124e-2,
2.0890373895444246e-2, 2.1613405906248049e-2, 2.2280051017541744e-2,
2.2888570026572980e-2, 2.3437375375404533e-2, 2.3925035292547804e-2,
2.4350277528205763e-2, 2.4711992673367795e-2, 2.5009237054089127e-2,
2.5241235193398702e-2, 2.5407381834409172e-2, 2.5507243519348632e-2};
/* A Legendre-Gaussian quadrature using 41 nodes which is exact for
a polynomial of up to order 59 */
```

```cpp
double PointF::intek(double ak,double bk,double(PointF::*grand)(double))
{
    int i;
    double sk,vk,ad,*pa,*pw;
    sk = bk - ak;
    vk = 0.025540559720393109*(this-->*grand)(ak+.5*sk);
    pa = add;
    pw = wkb;
    for(i=1;i<=GSQN;++i)
    {
        ad = sk**pa++;
        vk += ((this-->*grand)(ak+ad)+(this-->*grand)(bk-ad))**pw++;
    }
    return vk*sk;
}
// Integrand for the first part of the integral defined in [0,1]
double PointF::Fun1(double v)
{
    double t1,t2;
    t1 = v*(1.0-ao)/(1-ao*v);
    t2 = (1.0-ao*v)*(2.0*ao+Ga-3.0*Ga*v*ao);
    return t2*acos(t1)*FunS(ao*v);
}
// Integrand for the second part of the integral defined in [0,π/2]
double PointF::Fun2(double u)
{
    double su,t1,t2,t3;
    su = sin(u);
    t1 = 1.0-ao;
    t2 = t1+ao*su;
    t3 = t1*t1*(t1+(ao-Ga)*su)/(t2*t2*t2);
    return t3*FunS(ao*su/t2);
}
// Compute the integral for a given value of a
double PointF::FunF(double a)
{
    ao = a; // store the value of normalized crack size
    Ga = 3.0*(1.0-7.0*ao)*pow(1.0-ao,5.0)/28.0;
    double f1,f2;
    f = &PointF::Fun1;
    f1 = intek(0.0,1.0,f);
    f = &PointF::Fun2;
    f2 = intek(0.0,M_PI_2,f);
    return (f1+f2)/M_PI_2;
}
```

G.3 Sample Code for Usage of Class PointF

```cpp
#include ''stdafx.h"
#include <iostream>
#include <fstream>
#include <string>
#include <iomanip>
#include ''PointF.h"
int _tmain(int argc, _TCHAR* argv[])
{
    using namespace std;
    string str;
    size_t i,m;
    double x,x1,x2,y;
    double gl,dy;
    double thk;
// Create the class
    PointF Kf;

    cout << ''Output file name: ";
    cin >> str;
    ofstream ofile(str.c_str());
    if(!ofile)
    {
        cerr << ''Can't open input file ''" << str << "" " << endl;
        exit(EXIT_FAILURE);
    }
    cout << ''Number of data to be computed = ";
    cin >> m;
    cout << ''Starting and ending points on plane of cut = ";
    cin >> x1 >> x2;
    cout << ''Thickness = ";
    cin >> thk;
    cout << ''Gage length = ";
    cin >> gl;
    cout << ''Gage position offset = ";
    cin >> dy;
    if(dy>=gl/2)
    {
        cout << ''Error: Strain gage is off the centerline." << endl;
        exit(EXIT_FAILURE);
    }
// Compute the values of f^f(a,s)
    Kf.SetGPos(gl/thk,dy/thk);
    for(i=0;i<=m;i++)
    {
        x = (x1+(x2-x1)*i/m)/thk;
        y = Kf.FunF(x);
```

```
    ofile << std::fixed << x << '\t' << std::scientific
        << std::setprecision(10) << y << endl;
}
cout << ''Completed." << endl;
return 0;
}
```

References

1. K. E. Atkinson. An Intorduction to Numerical Analysis. John Wiley and Sons, New York, Chichester, Brisbane, Toronto, 1978.
2. C. C. Aydiner, E.Üstündag, M. B. Prime, and A. Peker. Modeling and measurement of residual stresses in bulk metallic glass plate. *Journal of Non-Crystalline Solids*, 316:82–95, 2003.
3. C. C. Aydiner and E. Üstündag. Residual stresses in a bulk metallic glass cylinder induced by thermal tempering. *Mechanics of Materials*, 37:201–212, 2004.
4. J. Lu A. Niku-Lari and J. F. Flavenot. Measurement of residual-stress distribution by the incremental hole-drilling method. *Experimental Mechanics*, 25:175–185, 1985.
5. N. P. Bansal and R. H. Doremus. *Handbook of Glass Properties*. Academic, Orlando, Florida, USA, 1986.
6. E. M. Beaney. Accurate measurement of residual stress on any steel using the centre hole method. *Strain*, 12:99, 1976.
7. M. Bijak-Zochowski. A semidestructive method of measuring residual stresses. *VID-Berichte*, 313:469–476, 1978.
8. E. Brinksmeier, J. T. Cammett, P. Leskovar W. Knig, J. Peters, and H. K. Tnshoff. Residual stresses - measurement and causes in machining processes. *Annals of CIRP*, 31:491–510, 1982.
9. H. F. Bueckner. Field singularity and integral expressions. In G. C. Sih, editor, *Methods of Analysis and Solutions of Crack Problems*, chapter 5, page 239. Noordhoff International publishing, Groningen, 1973.
10. G. Chell. The stress intensity factor for center and edge cracked sheets subjected to an arbitrary loading. *Fracture Mechanics*, 7:137–152, 1975.
11. Weili Cheng and Iain Finnie. A method for measurement of axisymmetric residual stresses in circumferentially welded thin-walled cylinders. *ASME Journal of Engineering Materials and Technology*, 106:181–185, 1985.
12. Weili Cheng and Iain Finnie. On the prediction of stress intensity factors for axisymmetric cracks in thin-walled cylinders from plane strain solutions. *ASME Journal of Engineering Materials and Technology*, 106:227–231, 1985.
13. Weili Cheng and Iain Finnie. Determination of stress intensity factors for partial penetration axial cracks in thin-walled cylinders. *ASME Journal of Engineering Materials and Technology*, 108:83–86, 1986.

14. Weili Cheng and Iain Finnie. Examination of the computational model of the layer removal method. *Experimental Mechanics*, 26:150–153, 1986.

15. Weili Cheng and Iain Finnie. Measurement of residual hoop stresses in cylinders using the compliance method. *ASME Journal of Engineering Materials and Technology*, 108:87–92, 1986.

16. Weili Cheng and Iain Finnie. A new method for measurement of residual axial stresses applied to a multi-pass butt-welded cylinder. *ASME Journal of Engineering Materials and Technology*, 109:337–342, 1987.

17. Weili Cheng and Iain Finnie. K_I solutions for an edge-cracked strip. *Eng. Fracture Mech.*, 31:201–207, 1988.

18. Weili Cheng and Iain Finnie. Stress intensity factors for radial cracks in cylinders and other simply closed bodies. *Eng. Fracture Mechanics*, 362:767–774, 1989.

19. Weili Cheng and Iain Finnie. The crack compliance method for residual stress measurement. *Welding in the World*, 28:103–110, 1990.

20. Weili Cheng and Iain Finnie. A K_{II} solution for an edge-cracked strip. *Eng. Fracture Mechanics*, 36:355–360, 1990.

21. Weili Cheng and Iain Finnie. An experimental method for determining residual stresses in welds. In M. Rappaz, editor, *Modeling of Casting, Welding and Advanced Solidification Processes - V*. The Minerals Metals and Materials Soc., 1991.

22. Weili Cheng and Iain Finnie. Measurement of residual stress distributions near the toe of a weld between a bracket and a plate using the crack compliance method. In L. Karisson, L. Lindgren, and M. Jonsson, editors, *Proc. of IUTAM Symposium, Mechanical Effects of Welding*, pages 135–141, Berlin Heidelberg, 1992. Springer-Verlag.

23. Weili Cheng and Iain Finnie. A prediction on the strength of glass following the formation of subsurface flaws by scribing. *J. of the American Ceramic Society*, 75:2565=–2572, 1992.

24. Weili Cheng and Iain Finnie. A comparison of the strains due to edge cracks and cuts of finite width with applications to residual stress measurement. *ASME Journal of Engineering Materials and Technology*, 115:220–226, 1993.

25. Weili Cheng and Iain Finnie. Measurement of residual stress distributions near the toe of an attachment welded to a plate using the crack compliance method. *Eng. Fracture Mechanics*, 46:79–92, 1993.

26. Weili Cheng and Iain Finnie. An overview of the crack compliance method for residual stress measurement. In *Proceedings of the Fourth International Conference on Residual Stresses*, pages 449–458, Baltimore, 1994.

27. Weili Cheng and Iain Finnie. Computation of stress intensity factors for a 2-d body from displacements at an arbitrary location. *Int. J. of Fracture*, 81:259–267, 1996.

28. Weili Cheng and Iain Finnie. Residual stress measurement by the introduction of slots or cracks. In Nisitani et al, editor, *Localized Damage IV*, pages 37–51. Computation Mechanics Publications, 1996.

29. Weili Cheng and Iain Finnie. The single-slice method for measurement of axisymmetric residual stresses in solid rods or hollow cylinders in the region of plane strain. *ASME Journal of Engineering Materials and Technology*, 120:170–176, 1998.

30. Weili Cheng. Measurement of the axial residual stresses using the initial strain approach. *ASME Journal of Engineering Materials and Technology*, 122:135–140, 2000.

31. W. Cheng, I. Finnie, M. Gremaud, A. Rosselet, and R.D. Streit. The compliance method for measurement of near surface residual stresses - application and validation for surface treatment by laser and shot-peening. *ASME Journal of Engineering Materials and Technology*, 116:556–560, 1994.

32. W. Cheng, I. Finnie, and Mike B. Prime. Measurement of residual stresses through the thickness of a strip using the crack-compliance method. In H. Fujiwara, T. Abe, and K. Tanaka, editors, *Residual Stresses - III Science and Technology*, pages 1127–1132, London and New York, 1992. Elsevier Science Publishers.

33. W. Cheng, I. Finnie, and R. O. Ritchie. Residual stress measurement on pyrolytic carbon-coated graphite leaflets for cardiac valve prostheses. In *Proc. 2001 SEM Annual Conference on Experimental and Applied Mechanics*, pages 604–607. Portland, Oregon, June 2001.

34. W. Cheng, I. Finnie, and Ö. Vardar. Measurement of residual stresses near the surface using the crack compliance method. *ASME Journal of Engineering Materials and Technology*, 113:199–204, 1990.

35. W. Cheng, I. Finnie, and Ö. Vardar. Deformation of an edge-cracked strip subjected to arbitrary shear surface traction on the crack faces. *Eng. Fracture Mechanics*, 43:33–40, 1992.

36. W. Cheng, I. Finnie, and Ö. Vardar. Deformation of an edge cracked strip subjected to normal surface traction on the crack faces. *Eng. Fracture Mech.*, 42:97–108, 1992.

37. W. Cheng, I. Finnie, and Ö. Vardar. Estimation of axisymmetric residual stresses in a long cylinder. *ASME Journal of Engineering Materials and Technology*, 114:137–140, 1992.

38. W. Cheng, M. Gremaud, M. Prime, and I. Finnie. Measurement of near surface residual stress using electric discharge wire machining. *ASME Journal of Engineering Materials and Technology*, 116:1–7, 1994.

39. W. Cheng, H. J. Schindler, and I. Finnie. Measurement of residual stress distribution in a disk or solid cylinder using the crack compliance technique. In *Proceedings of the Fourth International Conference on Residual Stresses*, Baltimore, 1994.

40. W. Cheng, G. Stevick, and I. Finnie. Prediction of the stress intensity factor for an internal circumferential crack at a butt-weld between cylinders using plane strain solutions. *ASME Journal of Engineering Materials and Technology*, 106:21–24, 1984.

41. J. W. Dally and W. F. Riley. *Experimental Stress Analysis*. MacGraw-Hill, Inc, N. Y.

42. Adrian T. DeWald and Michael R. Hill. Improved data reduction for the deep hole method of residual stress measurement. *J. of Strain Analysis for Engineering Design*, 38:65–78, 2003.

43. A.T. Dewald, J.E. Rankin, M.R. Hill, and K.I. Schaffers. An improved cutting plan for removing laser amplifier slabs from yb:s-fap single crystals using residual stress measurement and finite element modeling. *Journal of Crystal Growth*, 265:627–641, 2004.

44. A. T. DeWald, J. E. Rankin, M. R. Hill, M. J. Lee, and H. L. Chen. Assessment of tensile residual stress mitigation in alloy 22 welds due to laser peening. *ASME Journal of Engineering Materials and Technology*, 126:465–473, 2004.

45. R. R. de Swardt. Finite element simulation of crack compliance experiments to measure residual stresses in thick-walled cylinders. *J. Pressure Vessel Tech.*, 125:305–308, 2003.

46. C. Dalle Donne, G. Biallas, T. Ghidini, and G. Raimbeaux. Effect of weld imperfections and residual stresses on the fatigue crack propagation in friction stir welded joints. In *Proc. 2nd Int. Conf on Friction Stir Welding*, pages 26–28. Gothenberg Sweden, June 2000.

47. F. M. Ernsberger. Detection of strength-impairing surface flaws in glass. *Proc. Royal Soc.*, 257A:213–223, 1960.

48. N. Ersoy and Ö. Vardar. Measurement of residual stresses in layered composites by compliance method. *J. Composite Materials*, 34:575–598, 2000.

49. T. Fett and D. Munz. Proceedings 23, Vortragsveranstaltung des DVM-Arbetskreises Bruchvorgnge. Berlin, pages 249–259, 1991.

50. T. Fett. Bestimmung von eigenspannungen mittels bruchmechanischer beziehungen (determination of residual stresses by use of fracture mechanical relations). *Materialprfung*, 29:92–92, 1987.

51. Iain Finnie and Weili Cheng. Residual stresses and fracture mechanics. *ASME Journal of Engineering Materials and Technology*, 117:373–378, 1995.

52. Iain Finnie and Weili Cheng. A summary of past contributions on residual stresses. In *ECRS6 6th European Conf. on Residual Stresses*, pages 509–514, Coimbra, Portugal, 2002.

53. I. Finnie, W. Cheng, and K. J. McCorkindale. Delayed crack propagation in a steel pressure vessel due to thermal stresses. *Int. J. Pres. Ves. and Piping*, 42:15–31, 1990.

54. S. Finnie, W. Cheng, and I. Finnie. The computation and measurement of residual stresses in laser deposited layers. *ASME Journal of Engineering Materials and Technology*, 125:1–7, 2003.

55. M.T. Flaman. Brief investigation of induced drilling stresses in the center-hole method of residual-stress measurement. *Experimental Mechanics*, 1982.

56. Y. C. Fung. *Foundations of Solid Mechanics*. Prentice-Hall, Englewood Cliffs, N.J., 1965.

57. J.M. Ganley, A. K. Maji, and S. Huybrechts. Explaining spring-in in filament wound carbon fiber/epoxy composites. *J. Composite Materials*, 34:1216–1239, 2000.

58. C. J. Gilbert, V. Schroeder, and R. O. Ritchie. Mechanisms for fracture and fatigue-crack propagation in a bulk metallic glass. *Metall. & Mater. Trans. A*, 30:1739–1754, 1999.

59. R. D. Gregory. The spinning circular disc with a radial edge crack; an exact solution. *Int. J. of Fracture*, 41:39–50, 1989.

60. M. Gremaud, W. Cheng, I. Finnie, and M.B. Prime. The compliance method for measurement of near surface residual stresses - analytical background. *ASME Journal of Engineering Materials and Technology*, 116:550–555, 1994.

61. A. A. Griffith. The phenomena of rupture and flow in solids. *Phil Trans.*, 221A:163–198, 1921.

62. J. P. Den Hartog. *Advanced Strength of Materials*. McGraw-Hill, New York, 1952.

63. R. J. Hartrenft and G. G. Sih. Alternating method applied to edge and surface crack problems. In G. C. Sih, editor, *Methods of Analysis and Solutions of Crack Problems*, chapter 5, page 179. Noordhoff International publishing, 1973.

64. E Heyn and O. Bauer. ber spannungen in kaltgereckten metallen. *Int. Z. f. Metallographie*, 1:16, 1910.

65. Michael R. Hill and Wei-Yan Lin. Residual stress measurement in a ceramic-metallic graded material. *ASME Journal of Engineering Materials and Technology*, 124:185–191, 2001.

66. M. R. Hill and D. V. Nelson. Determining residual stress through the thickness of a welded plate. *ASME J. of PVP*, 318:343–352, 1996.

67. M. R. Hill and D. V. Nelson. Determining residual stress through the thickness of a welded plate. *ASME J. of PVP*, 327:29–36, 1997.

68. M. R. Hill and D. V. Nelson. The localized eigenstrain method for determination of triaxial residual stress in welds. *ASME J. of PVP*, 373:397–403, 1998.

69. G. R. Irwin. Fracture. In S. Flugge, editor, *Encycl. of Physics*, volume VI. Springer-Verlag, 1958.

70. D. E. Johnson and J. R. Johnson. *Mathematical Methods in Engineering and Physics*. Prentice-Hall, Inc, Englewood Cliffs, N. J., 1982.

71. M.R. Johnson, R.R. Robinson, A.J. Opinsky, M.W. Joerms, and D.H. Stone. Calculation of residual stresses in wheels from saw cut displacement data. Technical Report 85-WA/RT-16, 1985.

72. K.J. Kang, J.H. Song, and Y.Y. Earmme. A method for the measurement of residual stresses using a fracture mechanics approach. *Journal of Strain Analysis*, 24:23–30, 1989.

73. K. J. Kang, S. Darzens, and G. S. Choi. Effect of geometry and materials on residual stress measurement in thin films by using the focused ion beam. *ASME Journal of Engineering Materials and Technology*, 126:457–464, 2004.

74. D. A. Lados and D. Apelian. The effect of residual stress on the fatigue crack growth behavior of al-si-mg cast alloys - mechanisms and corrective mathematical models. *Metallurgical and Materials Transactions A*, 37A:133–145, 2006.

75. W. K. Lim, J. H. Song, , and Sankar B. V. Effect of ring indentation on fatigue crack growth in an aluminum plate. *Int. Journal of Fatigue*, 25:1271–1277, 2003.

76. J. Lu, editor. *Handbook of measurement of residual stresses*. Society for Experimental Mechanics, Lilburn, GA, 1996.

77. K. Masubushi. *Analysis of Welded Structures*. Pergamon Press, 1980.

78. J. Mathar. Determination of initial stresses by measuring the deformation around drilled holes. *Trans., ASME, Iron Steel*, 56(2):249–254, 1934.

79. R. Mayville and I. Finnie. Uniaxial stress-strain curves from a bending test. *Experimental Mechanics*, 22:197, 1982.

80. F. A. McClintock and A. S. Argon. *Mechanical Behavior of Materials*. Addision-Wesley, Reading, MA, 1966.

81. Measurement Group, Inc., Raleigh, NC. *Measurement of Residual Stresses by The Hole-Drilling Strain-Gage Method*, tech note tn503-2 edition, 1986.

82. M. Mesnager. Methode de determination des tensions existant dans un cylindre circulaire. *Academie des Sciences*, 169:1391, 1919.

83. C. R. Morin, R. J. Shipley, and J. A. Wilkinson. Fractography, nde, and fracture mechanics applications in failure studies. *Materials Characterization*, 33, 1994.

84. R. E. Mould. The strength and static fatigue of glass. *Glasstechn. Ber.*, III:18–28, 1959.

85. R. L. Martineau M. B. Prime. Mapping residual stresses after foreign object damage using the contour method. *Materials Science Forum*, 404-407:521–526, 2002.

86. H. Nisitani. Two-dimensional problem solved using a digital computer, 1967.

87. H. Nisitani. Solutions of notch problems by body force method. In G. C. Sih, editor, *Stress Analysis of Notch Problems, in Mechanics of Fracture*, 1978.

88. D. Nowell, S. Tochilin, and D. A. Hills. Use of the crack compliance method for the measurement of residual stress. In *Proc. of the 6th Int. Conf. on Residual Stresses*, volume 2, pages 845–852. Oxford, U.K., July 2000.

89. D. Nowell. Strain changes caused by finite width slots, with particular reference to residual stress measurement. *J. Strain Anal. Eng. Des.*, 34:285–294, 1999.

90. I. C. Noyan and J. B. Cohen. *Residual Stress Measurement by Diffraction and Interpretation*. MRE. New York, 1987.

91. R. Pathania, W. Cheng, M. Kreider, N. Cofie, and J. Brihmadesam. Analytical and experimental evaluation of residual stresses in bwr core shroud welds. In *ASME Press. Vess. and Piping Division Pub.*, volume PVP 373, pages 337–349. 1998.

92. H. J. Petroski and J. D. Achenbach. Computation of the weight function from a stress intensity factor. *Eng. Fracture Mechanics*, 10:257–266, 1978.

93. W. D. Pilkey. *Formulas for stress, strain and structural matrices*. Wiley, New York, 1994.

94. P. S. Prevey. Problems with non-destructive surface x-ray diffractiuon reisdual stress measurement. In C. Ruud, editor, *ASM Conf. Pro., Practical Applications of Residual Stress Technology*, pages 47–54, 1991.

95. M. B. Prime and I. Finnie. Surface strains due to face loading of a slot in a layered half-space. *ASME Journal of Engineering Materials and Technology*, 118:410–418, 1996.

96. M.B. Prime and Ch. Hellwig. Residual stresses in a bi-material laser clad measured using compliance. In T. Ericsson et al., editor, *The Fifth International Conference on Residual Stresses*, pages 127–132, Sweden, 1997. Linkping University.

97. M.B. Prime and M.R. Hill. Measurement of fiber-scale residual stress variation in a metal-matrix composite. *Journal of Composite Materials*, 38:2079–2095, 2004.

98. Mike B. Prime. Residual stress measurement by successive extension of a slot: The crack compliance method. *Applied Mechanics Reviews*, 52:75–96, 1999.

99. Mike B. Prime. Cross-sectional mapping of residual stresses by measuring the surface contour after a cut. *ASME Journal of Engineering Materials and Technology*, 123:162–168, 2001.

100. M. B. Prime and M. R. Hill. Uncertainty, model error, and order selection for series-expanded, residual-stress inverse solutions. *ASME Journal of Engineering Materials and Technology*, 128:175–185, 2006.

101. M. B. Prime. Experimental verification of the crack compliance method. M.s. project report, University of California, Berkeley, 1991.

102. E. Procter and B.M. Beaney. The trepan or ring-core method, centre-hole methods, sach's method and blind hole methods and deep hole technique. In *Advances in Surface Treament Technology, Vol. 4, Residual Stresses*, page 165. Permagon Press, New York, 1987.

103. I. S. Raju and Jr. J. C. Newman. Stress-intensity factors for a wide range of semi-elliptical surface cracks in finite-thickness plates. *Eng. Fracture Mechanics*, 11:817–829.

104. Jon E. Rankin, Michael R. Hill, and Lloyd A. Hackel. The effects of process variations on residual stress in laser peened 7049 t73 aluminum alloy. *Materials Science and Engineering A*, 349:279–291, 2003.

105. Jon E. Rankin and Michael R. Hill. Measurement of thickness-average residual stress near the edge of a thin laser peened strip. *ASME Journal of Engineering Materials and Technology*, 125:283–293, 2003.

106. C.N. Reid. A method of mapping residual stress in a compact tension specimen. *Scripta Metallurgica*, 22:451–456, 1988.

107. N. J. Rendler and I. Vigness. Hole-drilling strain gage method of measuring residual stresses. *Experimental Mechanics*, 6:577–586, 1966.

108. J. R. Rice. Some remarks on elastic crack-tip stress fields. *Int. J. Solid Structures*, 8:751–758, 1972.

109. D. Ritchie and R.H. Leggatt. The measurement of the distribution of residual stress through the thickness of a welded joint. *Strain*, pages 61–70, May 1987.

110. R. O. Ritchie. 1996.

111. D. P. Rooke and D. J. Cartwright. *Compendium of Stress Intensity Factors*. H.M.S.O., London, 1976.

112. E. F. Rybicki, J. R. Shadley, and W. S. Shealy. A consistent-splitting method for experimental residual-stress analysis. *Exp. Mech.*, 23:438–445, 1983.

113. G. Sachs. Der nachweis innerer spannungen in stangen und rohren. *Zeits Zeitshrift fr Metalkunde*, 19:352, 1927.

114. G. S. Schajer. Measurement of non-uniform residual stresses using the hole-drilling method. part *I*-stress calculation procedures. *ASME Journal of Engineering Materials and Technology*, 1103:338–343, 1988.

115. G. S. Schajer. Measurement of non-uniform residual stresses using the hole-drilling method. part ii-practical application of the integral method. *ASME Journal of Eng. Mat. and Tech.*, 110:344–349, 1988.

116. H.J. Schindler. Determination of residual-stress distributions from measured stress intensity factors. *Int. J. of Fracture*, 74:R23–R30, 1996.

117. H. J. Schindler, W. Cheng, and I. Finnie. Experimental determination of stress intensity factors due to residual stresses. *Experimental Mechanics*, 37:272–277, 1997.

118. H. J. Schindler. Weight functions for deep cracks and high stress gradients. In *ICF8*, Kiev, 1993.

119. H. J. Schindler. Residual stress measurement in cracked components: Capabilities and limitations of the cut compliance method. *Materials Science Forum*, 347-349:150–155, 2000.

120. G. S. Schjaer. Application of finite element calculation to residual stress measurements. *ASME Journal of Eng. Mat. and Tech.*, 1981.

121. W. Soete and R. Vancrombrugge. An industrial method for the determination of residual stresses. *Proceedings, SESA*, 8:17, 1950.

122. F. Stablein. Spannungsmessungen in eiseiting abgeloschten knüppeln. *Monatshefte*, 12:93–99, 1931.

123. H. Tada, P. Paris, and G. Irwin. *The Stress Analysis of Cracks Handbook*. Del Research Corperation, PA, 1973.

124. S. P. Timoshenko and J. N. Goodier. *Theory of Elasticity*. McGraw-Hill, New York, 1952.

125. S. Tochilin, D. Nowell, and D. A. Hills. Residual stress measurement by the crack compliance technique. In *Proc. 11th Int. Conf. on Exp. Mech.*, volume 2, pages 1313–1318. Oxford, UK, 1998.

126. R. G. Treuting and W. T. Read Jr. A mechanical determination of biaxial residual stress in sheet materials. *J. Appl. Phys.*, 22:131, 1951.

127. Murakawa H. Ueda, Y. and N. X. Ma. Measuring method for residual stresses in explosively clad plates and a method of residual stress reduction. *ASME Journal of Engineering Materials and Technology*, 118:576–582, 1996.

128. Y. Ueda, K. Fukuda, and Y. C. Kim. New measuring method of axisymmetric three-dimensional residual stresses using inherent strains as parameters. *ASME Journal of Engineering Materials and Technology*, 108:328–334, 1986.

129. Y. Ueda and K. Fukuda. New measuring method of three-dimensional residual stresses in long welded joints using inherent strains as parameters - lz method. *ASME Journal of Engineering Materials and Technology*, 111, 1989.

130. S. Vaidyanathan and I. Finnie. Determination of residual stresses from stress intensity factor measurements. *Journal of Basic Engineering*, 93:243–246, 1971.

131. S. M. Wiederhorn, H. Johnson, A. M. Diness, and A. H. Heuer. Fracture of glass in vacuum. *J. Am. Ceram. Soc.*, 57:336–341, 1974.

132. X. R. Wu and A. J. Carlsson. Weight functions and stress intensity factor solutions. 1991.

133. O. C. Zienkiewicz. *The Finite Element Method*. McGraw-Hill Book Company Limited, 3rd edition.

134. *E837-89 Standard Test Method for Determining Residual Stresses by the Hole-Drilling Strain-Gage Method*, 1989.

Index

Printed in the United States